D0399207

Lecture Notes in Physics

New Series m: Monographs

Springer
Berlin
Heidelberg
New York
Barcelona
Budapest
Hong Kong
London
Milan
Paris
Santa Clara
Singapore
Tokyo

The Editorial Policy for Monographs

The series Lecture Notes in Physics reports new developments in physical research and teaching - quickly, informally, and at a high level. The type of material considered for publication in the New Series m includes monographs presenting original research or new angles in a classical field. The timeliness of a manuscript is more important than its form, which may be preliminary or tentative. Manuscripts should be reasonably self-contained. They will often present not only results of the author(s) but also related work by other people and will provide sufficient motivation, examples, and applications.
The manuscripts or a detailed description thereof should be submitted either to one of the series editors or to the managing editor. The proposal is then carefully refereed. A final decision concerning publication can often only be made on the basis of the complete manuscript, but otherwise the editors will try to make a preliminary decision as definite as they can on the basis of the available information.
Manuscripts should be no less than 100 and preferably no more than 400 pages in length. Final manuscripts should preferably be in English, or possibly in French or German. They should include a table of contents and an informative introduction accessible also to readers not particularly familiar with the topic treated. Authors are free to use the material in other publications. However, if extensive use is made elsewhere, the publisher should be informed. Authors receive jointly 50 complimentary copies of their book. They are entitled to purchase further copies of their book at a reduced rate. As a rule no reprints of individual contributions can be supplied. No royalty is paid on Lecture Notes in Physics volumes. Commitment to publish is made by letter of interest rather than by signing a formal contract. Springer-Verlag secures the copyright for each volume.

The Production Process

The books are hardbound, and quality paper appropriate to the needs of the author(s) is used. Publication time is about ten weeks. More than twenty years of experience guarantee authors the best possible service. To reach the goal of rapid publication at a low price the technique of photographic reproduction from a camera-ready manuscript was chosen. This process shifts the main responsibility for the technical quality considerably from the publisher to the author. We therefore urge all authors to observe very carefully our guidelines for the preparation of camera-ready manuscripts, which we will supply on request. This applies especially to the quality of figures and halftones submitted for publication. Figures should be submitted as originals or glossy prints, as very often Xerox copies are not suitable for reproduction. For the same reason, any writing within figures should not be smaller than 2.5 mm. It might be useful to look at some of the volumes already published or, especially if some atypical text is planned, to write to the Physics Editorial Department of Springer-Verlag direct. This avoids mistakes and time-consuming correspondence during the production period.
As a special service, we offer free of charge LATEX and TEX macro packages to format the text according to Springer-Verlag's quality requirements. We strongly recommend authors to make use of this offer, as the result will be a book of considerably improved technical quality.
Manuscripts not meeting the technical standard of the series will have to be returned for improvement.
For further information please contact Springer-Verlag, Physics Editorial Department II, Tiergartenstrasse 17, D-69121 Heidelberg, Germany.

Mauro Ferrari Vladimir T. Granik
Ali Imam Joseph C. Nadeau (Eds.)

Advances in Doublet Mechanics

Springer

Editors

Mauro Ferrari
Vladimir T. Granik
Ali Imam
Joseph C. Nadeau
Department of Civil and Environmental Engineering
University of California at Berkeley
721 Davis Hall, Berkeley, CA 94720-1710, USA

Die Deutsche Bibliothek - CIP-Einheitsaufnahme

Advances in doublet mechanics / Mauro Ferrari ... (ed.). - Berlin ; Heidelberg ; New York ; Barcelona ; Budapest ; Hong Kong ; London ; Milan ; Paris ; Santa Clara ; Singapore ; Tokyo : Springer, 1997
(Lecture notes in physics : N.s. M, Monographs ; 45)
ISBN 3-540-62061-3
NE: Ferrari, Mauro [Hrsg.]; Lecture notes in physics / M

ISSN 0940-7677 (Lecture Notes in Physics. New Series m: Monographs)
ISBN 3-540-62061-3 Springer-Verlag Berlin Heidelberg New York

Typesetting: Camera-ready by editors/authors
Cover design: *design & production* GmbH, Heidelberg
SPIN: 10550845 55/3144-543210 - Printed on acid-free paper

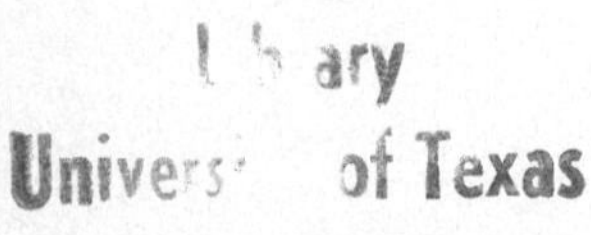

To Marialuisa Fusello Ferrari, *in memoriam*

To Serge Granik, *in memoriam*

To Federico Del Zotto, *in memoriam*

To my wife Paola, my son Giacomo, my daughters Chiara, Kim, Ilaria, and Federica, and my parents Antonio and Flavia. (MF)

To my wife, Galina. (VTG)

To my parents, Soraya and Zia. (AI)

To my parents, Roland and Suzanne. (JCN)

List of Contributors

Nasreen G. Chopra
Physics
University of California
Berkeley
CA 94720, USA

M. Ferrari
Civil and Environmental Engrg.
Materials Science and Mineral Engineering
Biomedical Microdevices Center
University of California
Berkeley
CA 94720-1710, USA

V. T. Granik
Civil and Environmental Engineering
University of California
Berkeley
CA 94720-1710, USA

A. Imam
Civil and Environmental Engineering
University of California
Berkeley
CA 94720-1710, USA

K. Mon
Materials Science and Mineral Engineering
University of California
Berkeley
CA 94720-1760, USA

J. C. Nadeau
Civil and Environmental Engineering
University of California
Berkeley
CA 94720-1710, USA

A. H. Nashat
Materials Science and Mineral Engineering
University of California
Berkeley
CA 94720-1760, USA

A. Zettl
Physics
University of California
Berkeley
CA 94720, USA

M. Zhang
Materials Science and Mineral Engineering
University of California
Berkeley
CA 94720-1760, USA

Preface

In 1991, my newly formed research group at Berkeley was working intensely in the area of continuum-level constitutive relationships that could be obtained in a deductive manner from microstructural information through the methods of homogenization theory. Of particular interest was the application of such methods to structural problems in the blossoming field of micromechanical devices. In this context it was becoming evident that we needed to learn to navigate through the continuum/discrete interface.

Such were the circumstances when Vladimir Granik came to visit us at Berkeley for the first time. It is probably not surprising that we received with great enthusiasm his offer to join forces and develop a mechanics of solid structures that would be based on a discrete representation of matter. Vladimir had established the foundations for such an endeavor with his work at Moscow University in the late 1970s.

Since that first meeting, and with ever-increasing enthusiasm, it has been a great privilege for me to collaborate with Vladimir. We first applied the formalism of what has become known as "doublet mechanics" to the microstructure-based theory of failure of solids and worked on the parallels and differences between the doublet approach and homogenization, together with Kevin Mon and Derek Hansford. Plane elastodynamics followed after Francesco Maddalena had proposed doublet viscoelesticity. The constitutive relationships in doublet mechanics were laid on a firm thermodynamical foundation through the work of Kevin Mon, while Miqin Zhang analyzed free boundary effects on multi-scale plane elastic waves in discrete domains. Joseph Nadeau and Amir Nashat made important contributions to plane elastostatics by establishing the doublet mechanical equivalent of the Airy's stress function method and obtaining the associated Green's function for isotropic domains.

The latter works are contained in this monograph, which was conceived so that a fairly complete, fully updated reference could be made available to researchers operating at the discrete/continuum interface. It is hoped that the methods of doublet mechanics will prove useful for further investigations in the many fields in which this interface is unavoidable. Among these are the science and technology of micromechanical devices, and of nanotubes, as discussed in Chap. 10 by Nasreen Chopra and Alex Zettl, as well as the mechanics of particulate and granular assemblies, the mechanics of soils and rocks, seismology, and the structure-property relationships of proteins, as discussed later in this book.

I wish to express my most heartfelt gratitude to all those who have contributed chapters to this volume. Special thanks to my co-editors Vladimir Granik, Ali Imam, and Joseph Nadeau for their careful planning and attentive review and integration of the contents. Further thanks to Joseph Nadeau for the vast majority of the technical aspects of the editorial work, and to Erin Cassidy for her assistance in this effort.

Finally, my warmest appreciation to Vladimir Granik, a great scientist—an even greater man.

Berkeley, CA
September 1996

Mauro Ferrari

Table of Contents

List of Figures

List of Tables

1. Introduction to Doublet Mechanics

M. Ferrari, V. T. Granik and A. Imam

1.1 Basic Concepts and Domains of Application

The mechanics of solids may be broadly divided into two major branches that differ on whether the underlying mass distribution is assumed to be discrete or continuous. Discrete models quite obviously reflect the nature of solids at the angstrom scale and may be supplemented with extremely detailed information on the interaction kinetics to yield predictive capabilities of extraordinary accuracy. continuum mechanics, on the other hand, is commonly based on the identification of actual media with regions of Euclidean space. In this domain, continuum mechanics has yielded results of exceptional beauty and very high practical value as a foundation for a multitude of engineering structural theories and computational schemes.

The most striking successes of discrete models have perhaps been reached under two very stringent conditions: the regularity of the underlying discrete assembly, forming a translational periodic arrangement or lattice; and the boundlessness of the material domain of interest. Thus, the dynamics of infinite single crystals of any material symmetry are very well understood. Considerable advances have been recorded in the mechanical modeling of individual defects in essentially infinite crystalline structures, such as point defects or dislocations. The interactions of pairs of defects have been successfully studied—with periodic arrays of defects—with the assumption of no interaction with real boundaries.

Yet the goal of employing discrete-based formalisms for the modeling of actual large-scale objects (say micron-sized or above) or structures of technological interest has been elusive. Actually, it has not been pursued by many in the scientific community, probably in view of the enormity of the computational resources required and the difficulty or impossibility to obtain the discrete-level or microstructural information required to implement a discrete model.

Of course, for very many situations of technological or engineering interest, there is no need whatsoever for a model based on the discrete nature of matter at the angstrom scale. Thus, continuum theories and their structural or computational derivatives have successfully answered many of the community's demands. However, the continuum approach certainly suffers from some fairly severe limitations of its own. In primis, in its most successful variant, due essentially to Cauchy, continuum mechanics is a non-scale theory, meaning that scaling effects cannot be accounted for. Of course, starting essentially in the 1960s, various elegant approaches such as those by Mindlin and Eringen have been proposed in order to inject scale- and microstructure-accounting features into the continuum framework. Yet it may be said that these generalized continuum mechanical theories have also imported analytical and computational difficulties that have discouraged their implementation outside of specialized academic circles.

Furthermore, all continuum theories are essentially deductive in nature: continuum-level governing variables are introduced, of both the kinetic and the kinematic types, and are then somehow identified with the physical reality to which they are designed to apply. Among such variables are the microstrains and microstresses of the micropolar approach and the director vectors of the directed-media approach to microstructural mechanics. Additional complications arise in conjunction with the model identification step of the micromechanical continuum approach: for instance, it is essentially impossible to relate quantities as basic as the Young's modulus or the yield stress of a polyphase, amorphous, or polycrystalline material to the microstructural properties of the material, be these the discrete-level or the phase-level physical properties, the properties of grain or phase boundaries, or the phase concentrations, orientation parameters, and other distributional data.

The great beauty of the continuum approach is, of course, that all of this micro-level information is very frequently irrelevant and may be bypassed altogether by identifying properties such as Young's modulus and yield stress directly from macroscopic experiments.

Yet there is a multitude of problems, of both academic and technological interest, which involve physical domains that are too irregular for application of continuum theories and yet too large, or again irregular, for modeling via the discrete approach. In a first approximation, these problems are defined by the fact that their material "microstructural" level is of dimensions comparable to the overall structural dimensions.

Problems of this nature arise in several, quite diverse disciplines. Microelectro-mechanical systems (MEMS), typically exhibiting dimensions in the micron-to-millimeter range, exhibit structural responses of strikingly different quality with respect to conventional engineering beams and plates, of which they are, after all, just a scaled-down version. The difference resides essentially in the fact that their constituent grains are also typically in the micron range and are thus comparable to typical cross-sectional dimensions. Along similar lines, buckeyballs and nanotubes are structural entities in their own right, yet they may not be modeled as conventional engineering structures in view of their evident discrete nature. The problem of protein unfolding, with the associated loss of protein function, is of essential mechanical nature. Yet it can not be addressed by a continuum model as a typical protein consists of a limited number of polypeptide chains. In geomechanics and seismology, great attention is dedicated to the so-called "granular" or "particulate" media, including sands and rock assemblies. Neglecting the discrete nature of these, as is done by resorting to a continuum model, results in paradoxical situations. This is exemplified by the so-called Flamant paradox (see Sect. 1.2) and the fact that most of seismology refers to non-scale elastic models of wave propagations. These models predict non-dispersive plane and surface waves, despite uniform experimental evidence to the contrary. In biomechanics, it is recognized that most biological structures present a multi-scale nature that may not be neglected. Among the outstanding examples of "granular" biological structures are the tissue of the heart and cancellous bone. Soft-tissue metastatic regions essentially present themselves as biphase assemblies of neoplastic cell clusters within a much softer interstitium. Modeling such regions as homogeneous continua leads to the neglecting of most significant phenomena, such as the rise of fluid pressure in the diseased region. This is known as the onset of "oncotic pressure" and is arguably the most important factor governing the penetration of chemotherapeutic agents within the cancer region.

The objective of this brief introduction is not a comprehensive listing of disciplines and problems for which a model bridging the discrete and the continuum approaches may be beneficial. Nor is it to review, critique, or compare the features of solid mechanics of the discrete and continuum type. The objective here is just to furnish motivation and interest in support of efforts in the development and application of models that would not be in contradiction with the great achievements in the fields of discrete and continuum mechanics, but which could contribute to bridging the gap between the two.

This monograph is dedicated to an effort of this type, and, in particular, to the theory of doublet mechanics (DM), which inherits its name from the fact that its "building block" is a pair of geometric points (nodes) at a finite distance, or a "doublet." In this sense, a doublet is to doublet mechanics what the "differential volume element" is to continuum mechanics. It is hoped that the material presented in this monograph will help establish that doublet mechanics is in full agreement with its two "boundary conditions," i.e. continuum mechanics on one extreme and lattice dynamics on the other. At the same time, it will be demonstrated that the theory is of truly multi-scale nature and may thus be employed for successfully addressing problems and resolving modeling paradoxes in very different disciplines, among which are those alluded to above. Finally, arguments will be presented to support the claims that microstructural identification in doublet mechanics is a realistic, and sometimes simple, objective.

Based on the original developments by Granik (1978), the theory of doublet mechanics was presented in a form that essentially referred to its linearly elastostatic geomechanical applications by Granik and Ferrari (1993). It was then extended to cover other domains of interest, including elastodynamics, viscoelasticity, failure theories, homogenization, and thermomechanics. An accounting of the state of doublet mechanics prior to this monograph is given in the next section, and an overview of the contents of the monograph is then provided in Sect. 1.3.

While a comprehensive review of the fundamental structure and equations of doublet mechanics is given in Sect. 1.4, at this point the basic concepts are summarized for the convenience of the reader.

1.1.1 Microstructure

In doublet mechanics, solids are represented as arrays of points or nodes at finite distances. Any pair of such nodes is termed a doublet. A bundle at a reference node is a collection of doublets involving the reference node. It must be clarified that the choice of what doublets will constitute a bundle is left open at the outset and may be different in different applications. Similarly, it is not necessary for the nodes to be periodically located, so as to form a lattice, even though it may be convenient to assume so for certain applications. If this assumption is introduced, then the bundle is typically chosen to correspond to the motif that generates the periodic structure. Other choices of bundles may reflect assumptions on the kinetics of doublet interaction. Thus, bundles may comprise all of the doublets (apart from symmetry) that interact with a given node. These may be the nearest neighbors, or may be defined by an assigned radius of influence, so that only the nodes with internodal distance less than such a radius will interact with the reference node.

If the choice is made that the nodes be located as to reproduce one of the fourteen Bravais lattices, then the typical bundle will consist of n basic (lattice) vectors, where $n = m/2$ is the valence of the lattice, and m is the coordination number of the array, i.e., the number of closest lattice points to any given lattice point. In such a setting, the internodal distances coincide with the lattice constants.

1.1.2 Interpretation

In the papers published prior to this monograph, the material model underlying doublet mechanics is a regular array of equal-sized elastic spheres of diameter d, the centers of which form a space Bravais lattice. This corresponds to the granular, or sphere-packing interpretation of doublet mechanics. The granular viewpoint offers ease of visualization, and on this basis it is frequently employed in the narrative of this monograph, and in particular in Sect. 1.4. Care must be exercised, however, not to confuse the visualization aid, that is the granular interpretation in this case, with the actual theory, which does not necessarily require such an interpretive aid.

Within this interpretation, which may be called *granular DM*, the following definitions are introduced:

The particle A is in contact with B_α if the doublet length η_α is equal to the true diameter of the sphere denoted by d. Such an α-doublet is called a *contact doublet*, the α-direction being the *contact direction*. There may be $s\,(0 \leq s \leq n)$ contact directions in which all particles are in contact, and the underlying granule array may be called a *regular s-contact n-valence array* H_{sn}. If the contact number $s = n$, then the array H_{sn} becomes a *regular completely contact* array $H_{nn} \equiv H_n$, or a *regular n-valence packing*, where all the adjacent particles are in contact with each other.

There are four regular packings (Deresiewicz 1958): H_3 (simple cubic), H_4 (cubical-tetrahedral), H_5 (tetragonal-sphenoidal), and H_6 (face-centered, or pyramidal). If the particle array H_{sn} is not a packing, then in addition to the packings mentioned above, there may be other regular structures similar to the crystal ones: simple tetragonal, orthorhombic face-centered, simple orthorhombic, etc. (Cottrell 1964). In these cases, the particles interact in the $n - s$ non-contact directions owing to intermediate substances (compliant inclusions) or spatial electrostatic forces binding atoms and molecules in crystals.

It should be noted that doublet micromechanics, as presented in Granik and Ferrari (1993), is valid for all the regular arrays H_{sn} whether $s = n$ or $s \neq n$, if the forces of particle interactions are of a short-distance character. Contact forces such as friction represent a particular case of such interactions.

The granular or sphere-packing interpretation of doublet mechanics is convenient, but by no means necessary. While the validity of doublet mechanics is apparent *a priori* for materials that exhibit a macroscopically evident granular lattice structure, it must be remarked that recent results—especially those concerning failure theories (see Sect. 1.3)—have established *a posteriori* its effectiveness for macroscopically continuous media. This suggests that doublet mechanics be interpreted as a general model, the validity of which is to be verified for specific material classes, much in analogy with what is current practice with continuum mechanics (CM). An interpretive aid in this context is the standpoint that the nodes of doublet mechanics actually be averages or representations of the complexity of actual particle or molecular interactions—a concept that is strongly related to the Clausius interpretation of the "molecular theory of elasticity" (see Todhunter (1886, Article No. 1400)).

It is now recalled that any arbitrary spatial distribution of points can be associated with a three-dimensional covering, consisting of the Voronoi cells including said points. The Voronoi cells for the Bravais lattices are regular polyhedra and are known as "Fedorov polyhedra." There are only five different Fedorov polyhedra (parallelohedra) for all of the Bravais lattices. On these premises, a different perspective on doublet mechanics may be constructed: doublet mechanics actually models solids as (space-filling) assemblies of Fedorov polyhedra. This corresponds to a representation of solid matter as a particulate assembly and gives rise to the so-called particulate interpretation of doublet mechanics.

1.1.3 Deformation

In doublet mechanics each lattice node is assumed endowed with a rotation and translation vector with increment vectors that may be expanded in a convergent Taylor series about the lattice nodal point. The order M at which the series is truncated defines the degree of approximation employed. The approximate theory with $M = 1$ does not contain any information on particle sizes or internodal distances and will be termed *nonscale*. Approximations with $M > 1$ include dependences on internodal distances.

In doublet mechanics nodes in any doublet may alter their axial distance, may rotate with respect to their common axis, and may relatively displace normally to their axis. Thus, doublet- or micro-strains of the longitudinal, torsional, and shear type are possible. Of these, the axial and torsional ones are *scalars* (not components of a tensor!), and the latter is a vector. Simple geometry permits the expressing of the microstrains via the displacement and rotation vectors and their derivatives.

Most importantly, the expressions relating microstrains to the displacement and rotation vectors also contain *the lattice geometry* and the *internodal distances* or particle dimensions. Based on this, doublet mechanics develops into a fully *multi-scale* theory. The first scale-related question that doublet mechanics helps answer is the following: under what ratios of the particle dimension to wavelength of the deformation is the use of the non-scale (continuum) kinematics justified? Granik and Ferrari (1993) have shown that the elongation microstrains can be determined with an error of 5% in the non-scale approximation (i.e., for the cited Taylor series truncated at $M = 1$) if the characteristic size D of the considered body domain is 27 times the particle diameter d. By contrast, for $D = 5d$, and $D = 2.5d$, the *scale* approximations corresponding to $M = 2$ and $M = 3$ are required to achieve the same lever of exactness. This proves that problems related to small body microdomains (stress concentration, propagation of waves of short wavelength, etc.) must be considered in the context of the novel micromechanical theory using the higher approximations.

1.1.4 Equilibrium

Microstresses of the axial, torsional, and shear type are introduced in doublet mechanics as energy conjugates of the corresponding microstrains. Energy methods are then employed to derive the microstress equations of conservation of linear and angular momenta. Most importantly, the natural boundary conditions descending from equilibrium in the integral energy formulation dictate the relationships between the microstresses and the macro- (or continuum level) stresses and stress couples. Macrostresses are expressed in terms of the microstresses, the lattice geometry, *and the typical internodal distance* with the desired degree of approximation.

On this basis, it will be demonstrated in Sect. 1.4 that the macrostresses are symmetric in a nonpolar medium with infinitesimal particles and are generally asymmetric for polar media. In addition to these physically expected results, doublet mechanics yields a novelty: the macrostresses are generally *asymmetric*, even in a *nonpolar* medium, if the scaling effects are not disregarded.

The relations between the micro- and the macro-stresses are employed to determine the macrostress equations of motion. In the non-scale, first approximation, these reduce to the classical ones for both the nonpolar (Cauchy) and the polar (Cosserat) media. *In the higher approximations, allowing for scaling effects, the equations of equilibrium have no counterpart in the standard mechanics of classical or microstructured continua.* For physical situations in which the nonscale theory is inappropriate (say, granular media traversed by high to medium relative frequency waves, in the sense specified above), even the use of polar continuum mechanics will yield erroneous results.

1.2 Previous Results in Doublet Mechanics

1.2.1 A Lesson Learned from Elastostatics

In Granik (1978) and Granik and Ferrari (1993), the fundamental system of equations, also to be presented in Sect. 1.4, was supplemented by linear elastic, homogeneous constitutive equations relating the microstresses to the microstrains. For the sake of illustration, the simplest variant of the doublet mechanical counterpart to the celebrated Flamant's problem was also addressed, whereby an isotropic, linear elastic semiplane was subjected to a point load on its free boundary. The nonscale variant of the problem was considered, under quasistatic, isothermal conditions, and with no body forces acting. Only longitudinal (nonpolar) and local interactions between the nodes were considered, corresponding to a medium that is characterized by one constitutive microconstant only.

Despite its simplicity, this problem yielded some qualitative results that were helpful in establishing the relationship between doublet mechanics and continuum mechanics: the microstresses were obtained in closed form and were found to be tensile in certain regions, even under a compressive load. It is recalled that the continuum-mechanical Flamant solution yields stresses that are compressive everywhere, in contrast with abundant experimental evidence reporting tensile openings in elastic granular media subjected to compressive boundary loads. Thus, doublet mechanical analysis of Flamant's problem provided a resolution of this conflicting state of affairs known as "Flamant's paradox."

The equilibrium relationships were then employed, to compute the macro- (continuum-level) stresses starting from the microstresses and the microstructural geometry. This yielded what was perhaps the most interesting conclusion; the continuum-level stresses were found to correspond exactly to Flamant's!

While not unexpected in terms of the general field structure of the theory, this provided a fairly immediate embodiment of a more general claim: doublet mechanics is in full agreement with the predictions of continuum mechanics, at least in the elastic domain, and yet there is significant, conceptually important conclusions that can be reached by doublet mechanical methods which are inaccessible by a continuum analysis.

The relationship between doublet and continuum mechanics requires a much more in-depth analysis than a simple example such as Flamant's may provide. The reader interested in a comparison between doublet mechanics and other field theories, such as continuum mechanics and lattice dynamics, is referred to the chapter by V. T. Granik, Chap. 4 of this monograph.

For completeness, it must be mentioned that the solution to the Flamant problem reported in Granik and Ferrari (1993) contained some algebraic errors that did not affect the nature of the findings contained therein. The correct solution is reported in the chapter by Nadeau, Nashat, and Ferrari, Chap. 8 of this monograph, which is dedicated to fundamental problems in doublet plane elastostatics.

1.2.2 Viscoelasticity

The general doublet mechanical linear viscoelastic constitutive relationships were introduced by Maddalena and Ferrari (1995). These relationships related microstresses to microstrains and directly closed the system of micro-level viscoelasticity equations. Constitutive expressions of the integral type were considered, in accordance with the so-called principle of fading memory, and micro-level quantities such as phase lag and loss modulus were identified.

Constitutive equations of the differential type were introduced in the same study and some specific rheological models were analyzed. In particular, the hypotheses of doublet interactions of the Maxwell (spring-dashpot in series) and Kelvin-Voigt (spring-dashpot in parallel) type were investigated.

The macroscopic counterpart to these doublet-level constitutive laws were then obtained by enforcing equilibrium requirements. It was determined that materials that are Kelvin-Voigt at the doublet level also present a continuum-level Kelvin-Voigt constitutive behavior. By contrast, it was only shown that strong conditions on the geometry of the underlying lattice would assure that doublet-level Maxwell materials would be of Maxwell type at the continuum level.

Equilibrium again allowed the formulation of the most general continuum-level viscoelastic constitutive relationship that is compatible with micro-level linear viscoelasticity. It was determined that such a macroconstitutive equation is characterized by the presence of projection tensors representing the spatial arrangement of the doublet vectors. To emphasize an important point: the continuum-level constitutive law was here derived exactly from a certain set of microstructural information. By contrast, typical continuum-level theories of the viscoelasticity of granular media introduce the notion of a constitutive "fabric tensor," and then impose restrictions on its possible dependences on the microstructural arrangements by means of conventional thermomechanics.

In a comparative analysis of the two approaches, the cited continuum fabric tensor viewpoint is perhaps found to be too general to be applicable in practice, as it will usually require a very high number of independent experiments in order to identify the multitude of physical constants it contains. The doublet mechanical viewpoint offers the advantages of identifying the possible form of the constitutive relationship and precisely indicating the dependance of the constitutive laws on the micro-level geometry and constitution. On the other hand, information on said micro-level quantities may be in itself extremely hard to obtain, thus limiting at times the practicality of the doublet approach as well.

Other results demonstrated in Maddalena and Ferrari (1995) include the fact that dissipation and material stability depend on the geometrical organization of the grains, and not only on their physical interactions as may be expected in a first approximation. Finally, the problem of the propagation of shear waves in a viscoelastic semispace was solved.

1.2.3 Failure Analysis

As was emphasized above, in doublet mechanics the transition from the doublet level to the continuum description is achieved in a natural manner via the enforcement of equilibrium requirements. This affords the definition of phenomenological variables such as the stresses in terms of basic microscopic level quantities. By enforcing equilibrium it was also possible to identify continuum-level elastic and viscoelastic constitutive tensors in terms of the doublet-level physical properties and the node arrangement.

A novel direction was investigated by Ferrari and Granik (1994, 1995) where the equilibrium-mediated micro-macro transition was employed in a derivation of continuum-level failure criteria starting from microstructural considerations.

The objective of these works was two-fold: to state and to characterize microstress-based (doublet level) failure criteria and to compare these criteria with their continuum-level counterparts. Of course, this comparison could only be performed in cases where there is a one-to-one correspondence between the doublet mechanical and the continuum mechanical treatments. For the case of interest this correspondence may be set up for conditions of plane stress. Thus, the treatment in both cases was restricted to plane conditions.

While on the topic of equivalence, a little digression is now proposed. As stated above, it is always possible to obtain the continuum-level stresses from the microstresses and microstructural information. However, it is not generally possible to obtain microstresses and microgeometry from the macrostresses alone. Herein lies the strength of the doublet mechanical approach: additional, micro-level information is made available that cannot be reached through conventional mechanics, while perfect harmony with the conventional macro-level treatment is maintained. In this sense, the micro-level stresses and deformations are the true fundamental variables of the theory. The conventional stresses are then second-order, phenomenological variables that become of importance only in the cases in which the underlying discrete nature of the material is completely irrelevant.

From the perspective of the proposal of new failure criteria important validating information could be gained from comparisons with the successfully established macro-level criteria such as Tresca's and Mises'. Thus, in Ferrari and Granik (1994), the first microstress-based criterion proposed was a polynomial of the second degree. Axial interactions only were considered, resulting in a three-parameter criterion, and an underlying hexagonal microstructure was assumed in order to ensure macroscopic in-plane nonscale isotropy. The three parameters were fixed by imposing limiting states in shear as well as tension and compression along two non-crystallographically equivalent directions. A novel criterion resulted that was expressed in terms of the axial microstresses.

The microstructural arrangement employed to derive this criterion was chosen specifically so that the macroscopic-level counterpart could be deduced. Thus, the microstresses were expressed in terms of the macrostresses and backsubstituted. The resulting, fully equivalent macrostress criterion was then found to coincide exactly with the established Goldenblatt-Copnov criterion for materials with different ultimate properties in tension and compression. If the tensile and compressive strengths coincide, then the criterion reduces to the von Mises plastification condition.

To the best of our knowledge, this was the first instance of a macroscopic level failure criterion deduced analytically from microscopic conditions. This may, in itself, be helpful as it offers insight on the assumptions implicitly contained in the stated criterion and, thus, on it limits of validity. However, the possible merits of the approach are not so much in its providing foundations for established results, but rather for the added ability to obtain more advanced, firmly founded failure criteria.

Thus, Ferrari and Granik (1994) introduced a new family of criteria based on the limit microstress concept, corresponding to the postulation that the limit macrostresses would be reached when all involved microstresses attain their limit values. In Ferrari and Granik (1995) the polynomial microstress-based failure criterion formulation was extended to third-degree polynomials. Again, attention was restricted to nonpolar, nonscale in-plane isotropic conditions. The family of resulting criteria contained four dimensionless parameters, and thus permitted the exact representation of five independent, macroscopic plane stress states. These were chosen to be uniaxial tension and compression, pure shear, and equibiaxial tension and compression. The novel criteria were shown to be successfully applicable to a variety of macroscopically continuous solids, including cast iron and several alloys, as well as to discontinua such as plain and short-steel-fiber reinforced concrete.

1.3 Overview of Contents

The governing equations expressing equilibrium and relating measures of deformation will be summarized in Sect. 1.4, following the developments originally presented in Granik (1978) and Granik and Ferrari (1993). Of course, these equations do not by themselves establish a well-posed mathematical problem and they need to be supplemented by constitutive equations expressing the physical properties of the media. A preliminary proposal for linear elastic constitutive laws was offered in the cited works. Its simplest subcase, corresponding to axial interactions only, was employed in all works on doublet elasticity predating this monograph.

The next two chapters of this volume, coauthored by Mon and Ferrari, are dedicated to establishing a firm foundation for the constitutive representation of elastic solids within the doublet mechanical framework. In particular, Chap. 2 sets forth the thermomechanical governing equations together with the methodology for employing them. The independent variables of the theory are the node displacement and rotation fields and temperature.

Following the approach of Green and Naghdi (1977), an entropy balance law is introduced, then reduced with the statement of local energy balance and finally employed as an identity. This entails stringent restrictions on the form of the possible constitutive dependences, and, in particular, rules out certain functional dependencies for the various thermomechanical variables. The energy-entropy balance-reduced elastic constitutive laws are further simplified by imposing a statement of the Second Law as embodied by the Clausius Inequality.

Rigid body motions for doublet mechanical assemblies are identified, and it is postulated that all constitutive laws be properly invariant under arbitrary rigid body motions superimposed on any given configuration. This yields additional restrictions on the functional dependencies. The chapter concludes with specification of the constitutive restrictions for the case of physically linear elastic response, and the formulation of the most general properly invariant and thermodynamically admissible micro-level linear elastic constitutive relations.

The developments of Chap. 2 require that the materials considered consist of a single component. It is recognized that a very wide variety of problems exist for which such an assumption is not tenable. Among these are typical materials science scenarios involving diffusion, precipitation, and segregation phenomena. On this basis, Chap. 3 addresses the problem of the elastic constitutive formulation for multi-component systems. The concept of mass balance is formulated and the requirements of thermodynamical admissibility and proper invariance under superposed rigid body motion are imposed. The treatment employs internal energy as thermodynamic potential, while in Chap. 2 the potential of choice was Helmholtz free energy.

From the review presentations of Sects. 1.1 and 1.2 it may be argued that doublet mechanics is in complete agreement with nonscale continuum mechanics, but is intrinsically richer, in that it yields information that is not accessible by conventional continuum mechanical means. In a much broader sense, involving the full multi-scale nature of the theory, the comparison of doublet mechanics with other approaches to the mechanics of solids is addressed by Granik in Chap. 4. It is recognized that the reader could have been curious about such comparison at earlier portions of the monograph, yet it is only at this point, after the development of fully invariant, admissible constitutive laws, that the theory is completely established and the comparison may be gainfully undertaken.

In particular, Chap. 4 briefly reviews some salient features of lattice dynamics, as well as continuum micromechanics or nonlocal continuum theories, of both the differential and the integral type. A comparison of these theories with doublet mechanics is then performed not only to address the formal differences, but, also, and perhaps most importantly, to identify the domains of application in which the use of doublet mechanics is expected to be advantageous.

In distinction to the first part of the monograph, Chaps. 5 through 10 are more applied in nature, in that they address the resolution of specific problems, or the application of the doublet mechanical formalism to specific disciplines.

In the first example of a fully multi-scale employment of doublet mechanics, the propagation of plane elastic waves is considered by Granik and Ferrari in Chap. 5. Here, the three-dimensional field equations of doublet elastodynamics are derived under some simplifying assumptions. The analysis is then focused on linear and planar arrangements, the latter with the further restriction of continuum-level planar isotropy.

Dispersion relations are established that demonstrate the dispersivity and retardation of both P- and S-waves at all scales other than those for which the continuum approximation is valid. The compatibility of the doublet mechanical analysis with lattice dynamics and continuum elasticity is demonstrated, and applications to crystals, granular media, and seismological problems are discussed.

The free-boundary reflection of plane waves in macroscopically isotropic solids is then addressed in Chap. 6 by Zhang and Ferrari. The effects of a typical microstructural dimensions on the reflection characteristics of P- and

S-waves are established. In particular, the scale-dependence of the critical angles of mode conversion, phase change and amplitude ratios are obtained. Of particular interest is the discovery that mode conversion takes place at different angles of incidence for materials with different microstructural dimensions (grain size). This may be employed as a basis for the determination of the microstructural scale that characterizes any microstructured medium of interest. Paralleling the previous chapters, the study by Zhang and Ferrari concludes with the derivation of the classical reflection results of continuum elastodynamics as a special non-scale case.

The analysis of elastodynamic phenomena is continued in Chap. 7 where Granik develops a new theoretical approach to the study of multi-scale surface waves in an elastic solid with cubic microstructure. A novel surface wave is identified that exhibits microscopic dispersion over relatively short wavelengths. In a departure from the previous chapters, difference equations are used in lieu of governing differential equations.

Nadeau, Nashat and Ferrari discuss macroscopically isotropic, non-scale plane elastostatics in Chap. 8. In this setting, a one-to-one correspondence between the continuum and the doublet variables may be established which permits the development of a technique whereby doublet mechanical solutions may be automatically generated from their continuum counterparts. By this method, micro-level stresses and deformations may be obtained directly from the wealth of established elasticity solutions. Thus, emphasis in problem-solving is shifted to the determination of the appropriate microstructural model for the medium of interest. In this sense, the sometimes formidable problems of mechanical analysis and microstructural identification are decoupled. The method is based on a novel uniqueness result for doublet elasticity.

In other developments in the chapter, a micro-stress function is introduced, in analogy with Airy's stress function in continuum elasticity. For illustration, several fundamental problems in doublet elasticity are solved, including Kelvin's and the problem of stress concentrations due to a circular hole.

Multi-scale doublet mechanical problems are amenable to a nested solution strategy whereby the nonscale solution is obtained first and then employed to generate a body-force-like term for the first-order scaling problem. The procedure may be inductively continued to higher-order scaling problems in a manner that is somewhat reminiscent of Signorini's method in nonlinear nonscale continuum elasticity. The described method of successive iterations is presented for the first time in Chap. 9 by Ferrari and Imam. An interesting result is that the nonscale solution of certain problems gives rise to a zero forcing term on the higher order problems. In these cases the nonscale solution is the exact solution and it may be concluded that the problem itself has a nonscale nature.

As discussed in Sect. 1.3, doublet mechanical analysis has been applied to crystalline solids, as well as granular, particulate, and composite media in the past. An exciting novel vista on the applications of the method is offered in Chap. 10 where Chopra and Zettl review the field of nanotubes. Nanotubes are technological objects of extraordinary interest. They are particularly suitable for doublet mechanical analysis because their discrete nature is evident at the structural scale and, as a result, they are not amenable to being modeled as continuous media.

1.4 Microstructure, Measures of Deformations, Field Equations

1.4.1 Kinematics

Microstucture. In this section, the fundamental equations of doublet mechanics are summarized. For ease of visualization, the granular interpretation of doublet mechanics is employed, whereby a solid is viewed as an assembly of equal spheres. The reader is reminded, however, that, as per the discussion in Sect. 1.1, such an interpretation is not necessary. Solids in doublet mechanics may equally well be modeled as space-filling assemblies of Fedorov polehedra, or as collections of regularly or irregularly placed nodes located at finite distances.

With this background, we consider a granular body as a spatial set H of a large number N of identical particles, viz., elastic spheres of diameter d. The particles are either in contact with each other or separated by adhesive layers.

To simplify the problem without losing its main features, we assume that the set $H_\circ$ (the subscript $\circ$ indicates the initial state of H) is regular: the centers of the particles, or nodes, form a Bravais lattice Γ.

A couple of adjacent particles A and B in $H_\circ$ represents a doublet (A, B) with the vector-axis $\boldsymbol{\zeta}^\circ$ emerging from the node $a \in A$ toward the node $b \in B$ (Fig. 1.1). The respective unit vector is $\boldsymbol{\tau}^\circ = \boldsymbol{\zeta}^\circ/\eta$ where $\eta \equiv \mid \boldsymbol{\zeta}^\circ \mid$ denotes the distance between a and b. If the particles are in contact then $\eta = d$.

Let $V_\circ$ denote the volume occupied by the set $H_\circ$. At every doublet (A, B_α) in $V_\circ$ we can attach an associated bundle $T_m(a)$ of m doublet axes $\boldsymbol{\zeta}_\alpha^\circ$ emerging from the node $a \in A$ toward all the adjacent nodes $b_\alpha \in B_\alpha$ (see Fig. 1.1). Here $\alpha \in \{1, 2, \ldots, m\}$; $m = 2n$; and n is the valence of the Bravais lattice.

For spherical particles whose nodes $a \in \Gamma$ the valence n ranges from 3 to 6. In particular, simple cubic (s.c.) and face-centered cubic (f.c.c.) structures have the valence $n = 3$ and $n = 6$, respectively (see Figs. 1.2 and 1.3). The FCCS is also called pyramidal.

For Bravais lattices the bundle $T_m(a)$ admits a decomposition into two disjoint subsets $T_n^+(a)$ and $T_n^-(a)$ which are equivalent via the center of

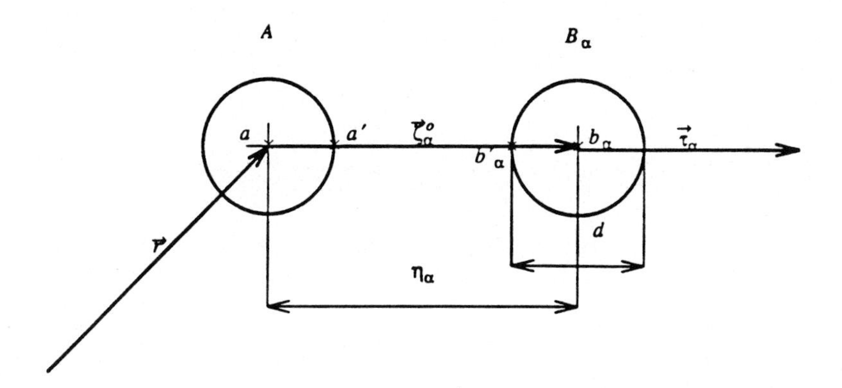

Fig. 1.1. Particle doublet (A, B_α).

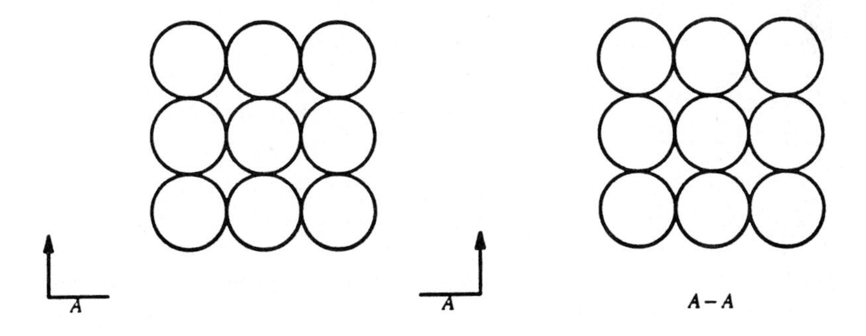

Fig. 1.2. Simple cubic structure (SCS).

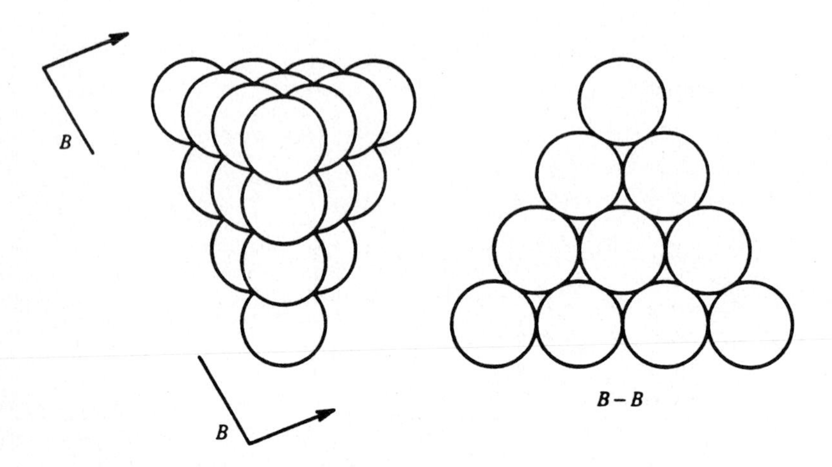

Fig. 1.3. Face-centered cubic structure (FCCS).

inversion operation at the node $a \in \Gamma$. In what follows it will suffice to consider only one of these subsets $T_n(a)$. The structural homogeneity of the lattice, for any fixed α, is represented by

$$\zeta_\alpha^\circ,\ \tau_\alpha^\circ = \text{constant} \qquad \forall a \in \Gamma. \tag{1.1}$$

Here, and in what follows, Greek subscripts (e.g., α, β) take on values in the range $1, \ldots, n$.

As the number of particles $N \to \infty$ and their diameter $d \to 0$, the volume of the primitive cell around each node approaches the elementary differential volume. This procedure allows one to replace summation over all the nodes $a \in \Gamma$ by integration over the volume $V_\circ$. Thus, if some function $\phi(a) = \sum_{\alpha=1}^{n} F_\alpha(a)$ is defined on a doublet bundle $T_n(a)$, then the transition to a continuous model is

$$\sum_{a\in\Gamma} \phi(a) = \sum_{a\in\Gamma}\sum_{\alpha=1}^{n} F_\alpha(a) = \int_{V_\circ} \sum_{\alpha=1}^{n} F_\alpha(\mathbf{x})\, dV, \tag{1.2}$$

where $\mathbf{x}$ is the position vector to the node $a \in \Gamma$ in $V_\circ$. Since the valence n of the Bravais lattice is constant we may interchange the summation and the integration equations (1.2) to yield

$$\sum_{a\in\Gamma}\sum_{\alpha=1}^{n} F_\alpha(a) = \sum_{\alpha=1}^{n} \int_{V_\circ} F_\alpha(\mathbf{x})\, dV. \tag{1.3}$$

This identity forms the basis for the subsequent transition to the continuous description of a discrete Bravais lattice Γ in the volume $V_\circ$ occupied by the granular body under consideration.

Microstrains. When a granular body undergoes deformation certain microstrains are developed in each of its doublets. We specify three of them:

- relative separation of the doublet nodes;
- rotation of the particles about the doublet axis;
- slipping of the particles past their point of contact.

The corresponding doublet microstrains are, respectively:

- elongation;
- torsion; and
- shear.

The above microstrains are induced by the translation of the nodes and by the rotation of the particles in the granular body. It should be noted that all these displacements are defined only at the nodes of a discrete Bravais lattice. The transition to a continuous description of the above displacements throughout the volume occupied by the granular body cannot be achieved exactly because the continuous volume is not isomorphic to any multitude of its points. Therefore, the problem of the transition to a continuum may be solved only in an approximate manner.

Approximate methods for performing such a transition were devised and applied to various microstructural models as reviewed, for example, by Teodorescu and Soós (1973). In the present work, we use a novel approach that is based on the following concept. We assume that the displacements of the particles vary little at the lengths on the order of their separations $\eta_\alpha \equiv \mid \zeta_\alpha^\circ \mid$. We then introduce two smooth mutually independent vector fields of the translations $\mathbf{u}(\mathbf{X}, t)$ and rotations $\boldsymbol{\phi}(\mathbf{X}, t)$, where $\mathbf{X}$ is a position vector of an arbitratry point in $V_\circ$ and t is time. We assume that these two vector fields coincide with the real translations and rotations of the granular body particles at the node $a \in \Gamma$, i.e., where $\mathbf{X} = \mathbf{x}$.

We also introduce two incremental vectors $\Delta \mathbf{u}_\alpha$ and $\Delta \boldsymbol{\phi}_\alpha$. The first of these is defined as

$$\Delta\, \mathbf{u}_\alpha \equiv \mathbf{u}(\mathbf{x} + \zeta_\alpha^\circ, t) - \mathbf{u}(\mathbf{x}, t). \tag{1.4}$$

It represents an increment of the translation vector $\mathbf{u}$ in a transition from an arbitrary node $a \in A$ to the adjacent node $b_\alpha \in B_\alpha$ (Fig. 1.4). The vector $\Delta \boldsymbol{\phi}_\alpha$ is defined analogously.

We assume that the above increment vectors may be expanded in a convergent Taylor series in a neighborhood of an arbitrary node $a \in \Gamma$ whose position vector is $\mathbf{x}$. Truncating this series at the M-th term we obtain

$$\Delta\, \mathbf{u}_\alpha = \sum_{\chi=1}^{M} \frac{(\eta_\alpha)^\chi}{\chi!} (\boldsymbol{\tau}_\alpha^\circ \cdot \nabla)^\chi\, \mathbf{u}(\mathbf{X}, t) \qquad (\text{when}\, \mathbf{X} = \mathbf{x}), \tag{1.5}$$

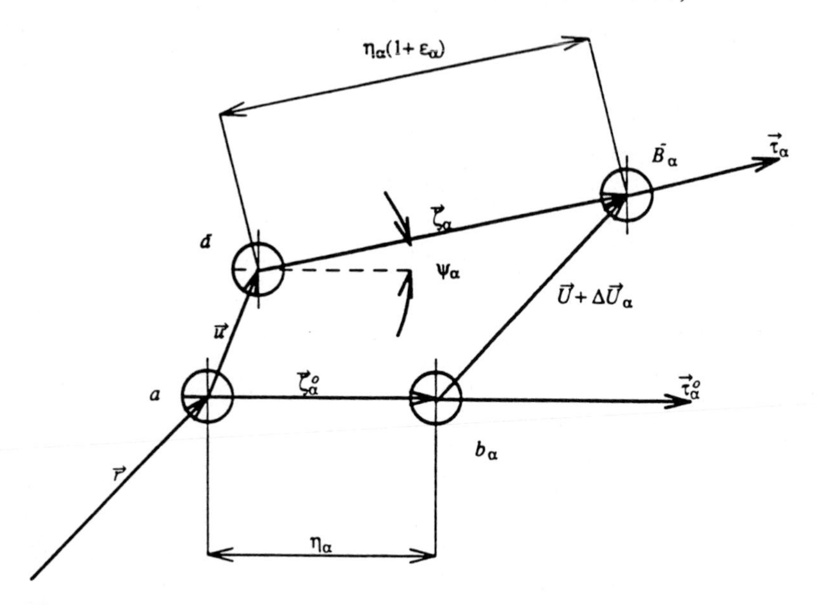

Fig. 1.4. Translations of the doublet nodes $a \in A$, $b_\alpha \in B_\alpha$.

where ∇ is the gradient operator and the dot denotes the inner product. The number M indicates the degree of approximation. The expansion for $\Delta\boldsymbol{\phi}_\alpha$ is defined analogously.

Furthermore, we introduce a stationary orthogonal Cartesian frame of reference x_i with a basis $\mathbf{e}_i$ $(i = 1,2,3)$. In this reference frame the above vectors and the operator ∇ are expressed as

$$\left.\begin{array}{lll} \boldsymbol{\tau}_\alpha^\circ = \tau_{\alpha i}^\circ \mathbf{e}_i & \mathbf{u} = u_i \mathbf{e}_i & \boldsymbol{\phi} = \phi_i \mathbf{e}_i \\ \mathbf{r} = x_i^\circ \mathbf{e}_i & \mathbf{R} = x_i \mathbf{e}_i & \nabla = \mathbf{e}_i \partial/\partial x_i . \end{array}\right\} \tag{1.6}$$

We adopt the convention that repeated Roman indices imply summation from 1 to 3. This convention does not apply to Greek subscripts.

In view of eqn. (1.6), the homogeneity condition takes the form

$$\tau_{\alpha i}^\circ = \text{constant}, \quad \forall\, \mathbf{x} \in V_\circ . \tag{1.7}$$

In order to derive the basic kinematic relations for the three microstrains, we consider an arbitrary doublet (A, B_α) with axis $\boldsymbol{\zeta}_\alpha^\circ$ in the initial region $V_\circ$. In the deformed region V of the current configuration this axis is mapped into the axis $\boldsymbol{\zeta}_\alpha$ (see Fig. 1.4), where

$$\boldsymbol{\zeta}_\alpha = \boldsymbol{\zeta}_\alpha^\circ + \Delta\mathbf{u}_\alpha . \tag{1.8}$$

According to eqn. (1.8), the corresponding director $\boldsymbol{\tau}_\alpha$ is

$$\boldsymbol{\tau}_\alpha \equiv \frac{\boldsymbol{\zeta}_\alpha}{\zeta_\alpha} = \frac{1}{1+\epsilon_\alpha}\left(\boldsymbol{\tau}_\alpha^\circ + \frac{\Delta\mathbf{u}_\alpha}{\eta_\alpha}\right), \tag{1.9}$$

where $\zeta_\alpha \equiv \mid \boldsymbol{\zeta}_\alpha \mid$, and

$$\epsilon_\alpha \equiv \frac{\zeta_\alpha - \eta_\alpha}{\eta_\alpha}. \tag{1.10}$$

Equation (1.10) represents the unit micro-elongation or elongation microstrain of an arbitrary doublet.

We assume that the relative displacements of the doublet nodes and the elongation microstrains are small, i.e., $| \Delta \mathbf{u}_\alpha | \ll \eta_\alpha$ and $\epsilon_\alpha \ll 1$, respectively. The approximate relation

$$\epsilon_\alpha = \frac{\boldsymbol{\tau}_\alpha^\circ \cdot \Delta \mathbf{u}_\alpha}{\eta_\alpha}, \tag{1.11}$$

then follows. Substituting $\Delta \mathbf{u}_\alpha$ from eqn. (1.5) into eqn. (1.11) and using eqns. (1.6) and (1.7), we find

$$\epsilon_\alpha = \tau_{\alpha i}^\circ \sum_{\chi=1}^{M} \frac{(\eta_\alpha)^{\chi-1}}{\chi!} \tau_{\alpha k_1}^\circ \cdots \tau_{\alpha k_\chi}^\circ \left. \frac{\partial^\chi u_i}{\partial x_{k_1} \cdots \partial x_{k_\chi}} \right|_{x=x^\circ} . \tag{1.12}$$

The equality $x = x^\circ$ indicates that after differentiation, the continuous coordinates $x_{k_1}, \ldots, x_{k_\chi}$ have to be replaced by discrete coordinates of the Bravais lattice nodes, i.e., $x_{k_1}^\circ, \ldots, x_{k_\chi}^\circ$. Each subscript of the set $k_1, \ldots, k_\chi$ runs through the integers $1, 2, 3$.

It should be noted that the elongation microstrain ϵ_α of the doublet (A, B_α) is caused by the motion of the node $b_\alpha \in B_\alpha$ away from node $a \in A$ along the vector $\boldsymbol{\tau}_\alpha$. Therefore, this microstrain can be conveniently represented as $\boldsymbol{\epsilon}_\alpha = \epsilon_\alpha \boldsymbol{\tau}_\alpha$. In the above discussion, it was assumed that $| \Delta \mathbf{u}_\alpha | \ll \eta_\alpha$ and $\epsilon_\alpha \ll 1$. Thus the angle ψ_α between the directors $\boldsymbol{\tau}_\alpha$ and $\boldsymbol{\tau}_\alpha^\circ$ is small, i.e., $\psi_\alpha \ll 1$ (see Fig. 1.4). Hence $\boldsymbol{\tau}_\alpha \approx \boldsymbol{\tau}_\alpha^\circ$ and we obtain the equality

$$\boldsymbol{\epsilon}_\alpha = \epsilon_\alpha \boldsymbol{\tau}_\alpha^\circ = \epsilon_\alpha \tau_{\alpha j}^\circ \, \mathbf{e}_j = \epsilon_{\alpha j} \, \mathbf{e}_j, \tag{1.13}$$

where $\epsilon_{\alpha j} \equiv \epsilon_\alpha \, \tau_{\alpha j}^\circ$.

It follows from eqn. (1.12) that the first approximation ($M = 1$) for the elongation microstrain has the form

$$\epsilon_\alpha = \tau_{\alpha i}^\circ \tau_{\alpha j}^\circ \, \varepsilon_{ij} \left.\right|_{x=x^\circ}, \tag{1.14}$$

where

$$\varepsilon_{ij} \equiv \frac{1}{2} \left(\frac{\partial u_i}{\partial x_j} + \frac{\partial u_j}{\partial x_i} \right) \tag{1.15}$$

are the components of the usual linear strain tensor. Expression (1.14) coincides with the one obtained earlier for the simplest model of granular media by Nikolaevskii and Afanasiev (1969).

Next, we examine the torsion microstrain. The torsion μ_α of the doublet (A, B_α) is caused by that part of the rotation vector increment $\Delta \boldsymbol{\phi}_\alpha$ which is directed along the unit vector $\boldsymbol{\tau}_\alpha^\circ$ (Fig. 1.5). By analogy with the microstrain of elongation, it then follows that

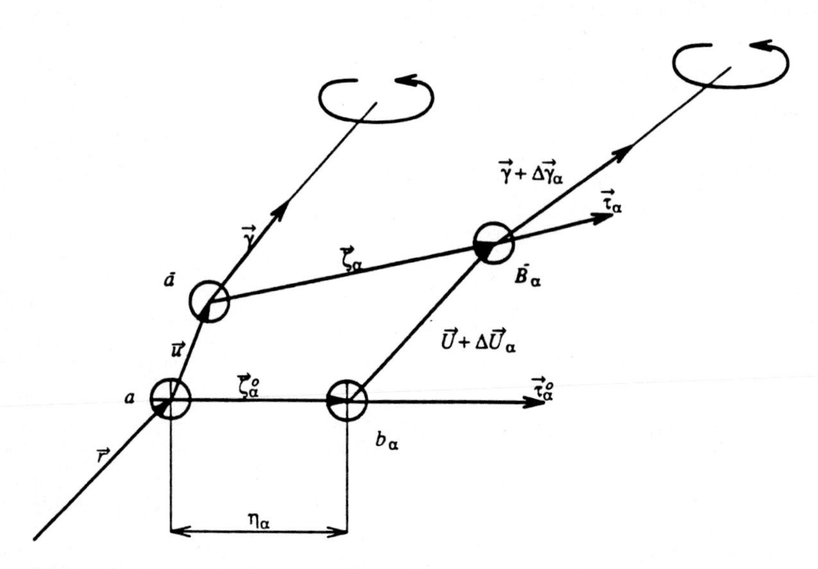

Fig. 1.5. Rotation of the doublet particles A and B_α.

$$\mu_\alpha = \frac{\boldsymbol{\tau}^\circ_\alpha \cdot \Delta\boldsymbol{\phi}_\alpha}{\eta_\alpha}. \tag{1.16}$$

Substituting the expansion of $\Delta\boldsymbol{\phi}_\alpha$, similar to eqn. (1.5), in eqn. (1.16) and taking into account eqns. (1.6) and (1.7), we find

$$\mu_\alpha = \tau^\circ_{\alpha i} \sum_{\chi=1}^{M} \frac{(\eta_\alpha)^{\chi-1}}{\chi!} \tau^\circ_{\alpha k_1} \ldots \tau^\circ_{\alpha k_\chi} \left. \frac{\partial^\chi \phi_i}{\partial x_{k_1} \ldots \partial x_{k_\chi}} \right|_{x=x^\circ}. \tag{1.17}$$

By analogy with eqn. (1.13), the torsion microstrain vector can be written as

$$\boldsymbol{\mu}_\alpha = \mu_\alpha \boldsymbol{\tau}^\circ_\alpha = \mu_\alpha \tau^\circ_{\alpha j} \mathbf{e}_j = \mu_{\alpha j} \mathbf{e}_j, \tag{1.18}$$

where $\mu_{\alpha j} \equiv \mu_\alpha \, \tau^\circ_{\alpha j}$.

Finally, we consider the shear microstrain. To this end, we note that the nodes a and b_α of an arbitrary doublet (A, B_α) have different displacements. Therefore, the doublet axis ζ°_α rotates by the angle $\boldsymbol{\psi}_\alpha$. Since, $| \Delta\mathbf{u}_\alpha | \ll 1$, it follows that $| \boldsymbol{\psi}_\alpha | \ll 1$ (see Fig. 1.4). If the independent rotations $\boldsymbol{\phi}$ of the granular body particles were not permitted, i.e., $\boldsymbol{\phi} \equiv 0$, the contact points $a' \in A$ and $b'_\alpha \in B_\alpha$ (see Fig. 1.1) would remain in contact after the deformation.

In the present theory such rotations are permitted. Owing to the independent rotations, $\boldsymbol{\phi} \neq 0$, the doublet particles A and B_α undergo, respectively, rotations by the small angles

$$\boldsymbol{\theta}_\alpha = \boldsymbol{\phi} - \boldsymbol{\psi}_\alpha, \quad \boldsymbol{\theta}'_\alpha = \boldsymbol{\theta}_\alpha + \Delta\boldsymbol{\phi}_\alpha. \tag{1.19}$$

Because of that the contact points move in opposite directions perpendicular to $\boldsymbol{\tau}^{\circ}_{\alpha}$. This leads to the shear of the doublet embodied in a shear microstrain.

The relative displacements of the contact points a' and b'_{α} are represented by the following cross products:

$$\begin{aligned}\Delta\,\mathbf{w} &= \frac{1}{2}\boldsymbol{\theta}_{\alpha}\times\boldsymbol{\zeta}^{\circ}_{\alpha} = \frac{1}{2}(\boldsymbol{\phi}-\boldsymbol{\psi}_{\alpha})\times\boldsymbol{\zeta}^{\circ}_{\alpha},\\ \Delta\,\mathbf{w}' &= -\frac{1}{2}\boldsymbol{\theta}'_{\alpha}\times\boldsymbol{\zeta}^{\circ}_{\alpha} = -\frac{1}{2}(\boldsymbol{\phi}+\Delta\boldsymbol{\phi}_{\alpha}-\boldsymbol{\psi}_{\alpha})\times\boldsymbol{\zeta}^{\circ}_{\alpha}.\end{aligned}\tag{1.20}$$

We assume here that the angle $\boldsymbol{\phi}$ is small so that $\mid\boldsymbol{\phi}\mid\ll 1$. The difference between $\Delta\mathbf{w}'$ and $\Delta\mathbf{w}$ divided by η_{α} yields the shear microstrain vector

$$\boldsymbol{\gamma}_{\alpha} = \boldsymbol{\psi}_{\alpha} - \boldsymbol{\phi} - \frac{1}{2}\,\Delta\,\boldsymbol{\phi}_{\alpha})\times\boldsymbol{\tau}^{\circ}_{\alpha}. \tag{1.21}$$

Since ψ_{α} is the small angle between the vectors $\boldsymbol{\tau}^{\circ}_{\alpha}$ and $\boldsymbol{\tau}_{\alpha}$ (see Fig. 1.4), the approximate relation

$$\boldsymbol{\psi}_{\alpha} = \boldsymbol{\tau}^{\circ}_{\alpha}\times\boldsymbol{\tau}_{\alpha} \tag{1.22}$$

holds.

Substituting the expansion for $\Delta\boldsymbol{\phi}_{\alpha}$, analogous to eqn. (1.5), along with $\boldsymbol{\phi}$ and $\boldsymbol{\psi}_{\alpha}$ from eqns. (1.6) and (1.22) into eqn. (1.21) and taking into account eqns. (1.7) and (1.9), we obtain

$$\boldsymbol{\gamma}_{\alpha} = \gamma_{\alpha i}\,\mathbf{e}_{i}, \tag{1.23}$$

where

$$\begin{aligned}\gamma_{\alpha i} &= -\left(\phi_j + \frac{1}{2}\sum_{\chi=1}^{M}\frac{(\eta_{\alpha})^{\chi}}{\chi!}\,\tau^{\circ}_{\alpha k_1}\ldots\tau^{\circ}_{\alpha k_{\chi}}\,\frac{\partial^{\chi}\phi_j}{\partial x_{k_1}\ldots\partial x_{k_{\chi}}}\bigg|_{x=x^{\circ}}\right)\tau^{\circ}_{\alpha p}\,\epsilon_{ijp}\\ &+ (\delta_{ij}-\tau^{\circ}_{\alpha i}\,\tau^{\circ}_{\alpha j})\sum_{\chi=1}^{M}\frac{(\eta_{\alpha})^{\chi-1}}{\chi!}\,\tau^{\circ}_{\alpha k_1}\cdots\tau^{\circ}_{\alpha k_{\chi}}\,\frac{\partial^{\chi}u_j}{\partial x_{k_1}\ldots\partial x_{k_{\chi}}}\bigg|_{x=x^{\circ}}\end{aligned}\tag{1.24}$$

and where ϵ_{ijp} is the permutation symbol and δ_{ij} is the Kronecker delta.

1.4.2 Microstresses, Equations of Motion, Boundary Conditions

We postulate the existence of the following internal generalized microstresses associated with the internal generalized microstrains.

– The elongation microstresses $\mathbf{p}_{\alpha}$ associated with $\boldsymbol{\epsilon}_{\alpha}$:

$$\mathbf{p}_{\alpha} = p_{\alpha}\,\boldsymbol{\tau}^{\circ}_{\alpha} = p_{\alpha}\,\tau^{\circ}_{\alpha i}\,\mathbf{e}_i = p_{\alpha i}\,\mathbf{e}_i. \tag{1.25}$$

– The torsional microstresses $\mathbf{m}_{\alpha}$ associated with $\boldsymbol{\mu}_{\alpha}$:

$$\mathbf{m}_{\alpha} = m_{\alpha}\,\boldsymbol{\tau}^{\circ}_{\alpha} = m_{\alpha}\,\tau^{\circ}_{\alpha i}\,\mathbf{e}_i = m_{\alpha i}\,\mathbf{e}_i. \tag{1.26}$$

– The shear microstresses $\mathbf{t}_{\alpha}$ associated with $\boldsymbol{\gamma}_{\alpha}$:

$$\mathbf{t}_{\alpha} = t_{\alpha i}\,\mathbf{e}_i. \tag{1.27}$$

If the strain energy per unit volume of the elastic granular medium is denoted by $W = W(\epsilon_\alpha, \mu_\alpha, \gamma_{\alpha i})$, then the stress power P is given by

$$P = \dot{W} = \sum_{\alpha=1}^{n} (p_\alpha \, \dot{\epsilon}_\alpha + m_\alpha \dot{\mu}_\alpha + t_{\alpha i} \, \dot{\gamma}_{\alpha i}) . \tag{1.28}$$

On the other hand, the time rate of change of W is

$$\dot{W} = \sum_{\alpha=1}^{n} \left(\frac{\partial W}{\partial \epsilon_\alpha} \dot{\epsilon}_\alpha + \frac{\partial W}{\partial \mu_\alpha} \dot{\mu}_\alpha + \frac{\partial W}{\partial \gamma_{\alpha i}} \dot{\gamma}_{\alpha i} \right) . \tag{1.29}$$

Comparing eqns. (1.28) and (1.29) we find

$$p_\alpha = \frac{\partial W}{\partial \epsilon_\alpha}, \qquad m_\alpha = \frac{\partial W}{\partial \mu_\alpha}, \qquad t_{\alpha i} = \frac{\partial W}{\partial \gamma_{\alpha i}} . \tag{1.30}$$

We also define the functional

$$\begin{aligned} L[\mathbf{u}, \phi] &= \int_0^T \int_V \left(W - \frac{\rho}{2} \mid \dot{\mathbf{u}} \mid^2 - \mathbf{F} \cdot \mathbf{u} \right) dv \, dt \\ &\quad - \int_0^T \int_S (\mathbf{T} \cdot \mathbf{u} + \mathbf{M} \cdot \phi) \, ds \, dt \end{aligned} \tag{1.31}$$

for the granular medium, where $\mathbf{T}$ and $\mathbf{M}$ are the force and couple vectors per unit area of the surface S, respectively, $\mathbf{F}$ is the force per unit volume and ρ is the density.

Taking the first variation of L yields

$$\left. \frac{d}{d\lambda} L[\mathbf{u} + \lambda \mathbf{v}, \phi + \lambda \boldsymbol{\theta}] \right|_{\lambda=0} , \tag{1.32}$$

where $\mathbf{v}$ and $\boldsymbol{\theta}$ are arbitrary smooth functions with $\mathbf{v}(\mathbf{R}, 0) = \mathbf{v}(\mathbf{R}, T) = 0$, and setting it equal to zero and using eqn. (1.30), along with the definitions of the microstrains, yield the following results in terms of the microstresses:

1) Conservation of linear momentum:

$$\begin{aligned} &\sum_{\alpha=1}^{n} \sum_{\chi=1}^{M} (-1)^{\chi-1} \frac{(\eta_\alpha)^{\chi-1}}{\chi!} \overset{\circ}{\tau}_{\alpha k_1} \cdots \overset{\circ}{\tau}_{\alpha k_\chi} \frac{\partial^\chi (t_{\alpha i} + p_{\alpha i})}{\partial x_{k_1} \ldots \partial x_{k_\chi}} + F_i \\ &= \rho \frac{\partial^2 u_i}{\partial t^2} , \end{aligned} \tag{1.33}$$

conservation of moment of momentum:

$$\begin{aligned} &\sum_{\alpha=1}^{n} \left(\epsilon_{ijq} \overset{\circ}{\tau}_{\alpha j} t_{\alpha q} + \sum_{\chi=1}^{M} (-1)^{\chi-1} \frac{(\eta_\alpha)^{\chi-1}}{\chi!} \overset{\circ}{\tau}_{\alpha k_1} \cdots \overset{\circ}{\tau}_{\alpha k_\chi} \right. \\ &\quad \left. \times \frac{\partial^\chi (m_{\alpha i} - \frac{1}{2} \eta_\alpha \, \epsilon_{ijq} \overset{\circ}{\tau}_{\alpha j} t_{\alpha q})}{\partial x_{k_1} \ldots \partial x_{k_\chi}} \right) = 0 . \end{aligned} \tag{1.34}$$

2) Natural boundary conditions at the surface S
force boundary conditions:

$$n_{k_r} \sum_{\alpha=1}^{n} \overset{\circ}{\tau}_{\alpha k_r} \sum_{\chi=r}^{M} (-1)^{\chi-1} \frac{(\eta_\alpha)^{\chi-1}}{\chi!} \overset{\circ}{\tau}_{\alpha k_{r+1}} \cdots \overset{\circ}{\tau}_{\alpha k_\chi} \frac{\partial^{\chi-r}(t_{\alpha i} + p_{\alpha i})}{\partial x_{k_{r+1}} \cdots \partial x_{k_\chi}} = T_i \, \delta_{r1}, \tag{1.35}$$

couple boundary conditions:

$$n_{k_r} \sum_{\alpha=1}^{n} \overset{\circ}{\tau}_{\alpha k_r} \sum_{\chi=r}^{M} (-1)^{\chi} \frac{(\eta_\alpha)^{\chi-1}}{\chi!} \overset{\circ}{\tau}_{\alpha k_{r+1}} \cdots \overset{\circ}{\tau}_{\alpha k_\chi} \times \frac{\partial^{\chi-r}(m_{\alpha i} - \frac{1}{2}\eta_\alpha \, \epsilon_{ijq} \, \overset{\circ}{\tau}_{\alpha j} \, t_{\alpha q})}{\partial x_{k_{r+1}} \cdots \partial x_{k_\chi}} = - M_i \, \delta_{r1}. \tag{1.36}$$

Here $\mathbf{n}$ denotes the outward unit normal to the surface S. The subscript $r = 1, \ldots, M-1$ if $M \geq 2$, and $r = 1$ if $M = 1$. However, r should be such that $r + 1 \leq \chi$ otherwise one has to take the product $\overset{\circ}{\tau}_{\alpha k_{r+1}} \cdots \overset{\circ}{\tau}_{\alpha k_\chi} = 1$ and to ignore the differential operator in eqns. (1.35) and (1.36) so that $\partial^{\chi-r}(\cdots) = (\cdots)$.

To clarify the indicial convention, let $M = 3$. Then the maximum value of r is $r = M - 1 = 2$, and eqn. (1.35) gives

(a) for $r = 1$:

$$n_{k_1} \sum_{\alpha=1}^{n} \overset{\circ}{\tau}_{\alpha k_1} \left[(t_{\alpha i} + p_{\alpha i}) - \frac{\eta_\alpha}{2!} \overset{\circ}{\tau}_{\alpha k_2} \frac{\partial (t_{\alpha i} + p_{\alpha i})}{\partial x_{k_2}} + \frac{(\eta_\alpha)^2}{3!} \overset{\circ}{\tau}_{\alpha k_2} \overset{\circ}{\tau}_{\alpha k_3} \frac{\partial^2 (t_{\alpha i} + p_{\alpha i})}{\partial x_{k_2} \partial x_{k_3}} \right] = T_i, \tag{1.37}$$

(b) for $r = 2$:

$$n_{k_2} \sum_{\alpha=1}^{n} \overset{\circ}{\tau}_{\alpha k_2} \left[\frac{\eta_\alpha}{2!} (t_{\alpha i} + p_{\alpha i}) - \frac{(\eta_\alpha)^2}{3!} \overset{\circ}{\tau}_{\alpha k_3} \frac{\partial (t_{\alpha i} + p_{\alpha i})}{\partial x_{k_3}} \right] = 0. \tag{1.38}$$

1.4.3 Transition from Microstresses to Macrostresses

We represent the components of the surface force vector $\mathbf{T}$ and couple vector $\mathbf{M}$ in the form

$$T_i = \sigma_{ki} \, n_k, \qquad M_i = M_{ki} \, n_k, \tag{1.39}$$

where σ_{ki} and M_{ki} are components of the second order tensors of the force macrostresses $\mathbf{T}$ and the couple macrostresses $\mathbf{M}$, respectively. By comparing eqn. (1.39) to eqns. (1.35) and (1.36), we easily find the natural connection between the micro- and the macrostresses:

$$\sigma_{k_1 i}^{(M)} = \sum_{\alpha=1}^{n} \overset{\circ}{\tau}_{\alpha k_1} \sum_{\chi=1}^{M} (-1)^{\chi+1} \frac{(\eta_\alpha)^{\chi-1}}{\chi!} \overset{\circ}{\tau}_{\alpha k_2} \cdots \overset{\circ}{\tau}_{\alpha k_\chi} \frac{\partial^{\chi-1}(t_{\alpha i} + p_{\alpha i})}{\partial x_{k_2} \ldots \partial x_{k_\chi}}, \quad (1.40)$$

$$\begin{aligned} M_{k_1 i}^{(M)} &= \sum_{\alpha=1}^{n} \overset{\circ}{\tau}_{\alpha k_1} \sum_{\chi=1}^{M} (-1)^{\chi+1} \left[\frac{(\eta_\alpha)^{\chi-1}}{\chi!} \overset{\circ}{\tau}_{\alpha k_2} \cdots \overset{\circ}{\tau}_{\alpha k_\chi} \right. \\ &\quad \left. \times \frac{\partial^{\chi-1}(m_{\alpha i} - \frac{1}{2}\eta_\alpha \, \epsilon_{ijq} \, \overset{\circ}{\tau}_{\alpha j} \, t_{\alpha q})}{\partial x_{k_2} \ldots \partial x_{k_\chi}} \right]. \end{aligned} \quad (1.41)$$

The superscript M gives the level of the approximation at which the macrostresses are represented by the microstresses.

A brief digression is now undertaken in order to discuss the symmetry of the macroscopic stress tensor accounting for its scale dependence. Toward this objective a simplified system of equations is derived for the special case corresponding to the assumptions that $p_{\alpha i} = 0$, $m_{\alpha i} = 0$ (doublet-level nonpolar medium) and $t_{\alpha i}$ depend only on x_3. Furthermore, a simplified underlying microstructural geometry is assumed whereby $\overset{\circ}{\tau}_{\alpha i} = 1$ if $\alpha = i$, $\overset{\circ}{\tau}_{\alpha i} = 0$ otherwise, and $\eta_\alpha = \eta > 0$ for $\alpha = 1, 2, 3$.

With these assumptions, the equation of balance of linear momentum (1.33) reduces to

$$t_{3i,3} - \frac{\eta}{2} t_{3i,33} = 0. \quad (1.42)$$

These can be integrated to yield

$$t_{3i} = A_0 + A_i \exp(kx_3), \quad (1.43)$$

where $k = 2/\eta$, and A_0 and A_i $(i = 1, 2, 3)$ are arbitrary constants.

The equation of balance of angular momentum (1.34) reduces to

$$\epsilon_{ijq} \, t_{jq} - \frac{\eta}{2} \epsilon_{i3q} \, t_{3q,3} + \frac{\eta^2}{4} \epsilon_{i3q} \, t_{3q,33} = 0. \quad (1.44)$$

These yield the conditions

$$t_{23} = t_{32}, \qquad t_{13} = t_{31}, \qquad t_{12} = t_{21}. \quad (1.45)$$

On the basis of these results, it can be established that the macroscopic stress tensor is asymmetric. For instance, provided $A_2 \neq 0$, it follows that

$$\sigma_{32}^{(2)} - \sigma_{23}^{(2)} = -\frac{\eta}{2} t_{32,3} = -\frac{\eta}{2} A_2 \exp(kx_3) \neq 0. \quad (1.46)$$

Thus, if scale effects are included, the macroscopic stress tensor need not be symmetric—even for nonpolar media.

Let us return to eqns. (1.40) and (1.41) having another goal in mind. By comparing these with eqns. (1.33) and (1.34), we find the macrostress equations of motion for the volume V of the granular medium to be

$$\frac{\partial \sigma_{ij}^{(M)}}{\partial x_i} + F_j = \rho \frac{\partial^2 u_j}{\partial t^2}, \tag{1.47}$$

$$\frac{\partial M_{ij}^{(M)}}{\partial x_i} + \epsilon_{jik}\, \sigma_{ik}^{(1)} = 0. \tag{1.48}$$

Taking into account eqns. (1.40) and (1.41), we can represent the boundary conditions (1.35) and (1.36) for the macrostresses on the surface S of the granular body as

$$T_j = \sigma_{ij}^{(M)}\, n_i, \qquad M_j = M_{ij}^{(M)}\, n_i. \tag{1.49}$$

In the first approximation ($M = 1$) eqns. (1.47) and (1.48) describe the motion of Cosserat's continuum with asymmetric force macrostresses (Cosserat and Cosserat 1909). Under some conditions (see, below, Sect. 4.3.2 the couple macrostresses M_{ij} vanish. The relations (1.48) then bring about symmetric force macrostresses and thus reduce the relations (1.47) to the equations of motion of the "symmetric" classical continuum. In this case, the couple boundary conditions $(1.49)_2$ become identities.

In the subsequent approximations ($M = 2, 3, ...$) eqns. (1.47) and (1.48) include different force macrostresses: $\sigma_{ij}^{(M)}$ in eqn. (1.47) and $\sigma_{ij}^{(1)}$ in eqn. (1.48). Due to this feature, the macroscopic equations (1.47) and (1.48) at $M > 1$ become quite different from the equations of motion of both classical and Cosserat continua.

1.4.4 Alternative Formulation

For later convenience, we express some of the equations developed in the previous sections in a more compact form. The kinematic equations (1.12), (1.17) and (1.24) are written as

$$\epsilon_\alpha = \sum_{\chi=1}^{M} \frac{(\eta_\alpha)^{\chi-1}}{\chi!}\, \overset{\circ}{\tau}_{\alpha i}\, T_{\alpha(\chi)}\, \partial^\chi u_i, \tag{1.50}$$

$$\mu_\alpha = \sum_{\chi=1}^{M} \frac{(\eta_\alpha)^{\chi-1}}{\chi!}\, \overset{\circ}{\tau}_{\alpha i}\, T_{\alpha(\chi)}\, \partial^\chi \phi_i, \tag{1.51}$$

$$\begin{aligned}\gamma_{\alpha i} = {} & (\delta_{ij} - \overset{\circ}{\tau}_{\alpha i}\, \overset{\circ}{\tau}_{\alpha j}) \sum_{\chi=1}^{M} \frac{(\eta_\alpha)^{\chi-1}}{\chi!}\, T_{\alpha(\chi)}\, \partial^\chi u_j \\ & - \left(\phi_i + \frac{1}{2} \sum_{\chi=1}^{M} \frac{(\eta_\alpha)^{\chi}}{\chi!}\, T_{\alpha(\chi)}\, \phi_j \right) \overset{\circ}{\tau}_{\alpha p}\, \epsilon_{ijp},\end{aligned} \tag{1.52}$$

where

$$T_{\alpha(\chi)} \equiv \mathring{\tau}_{\alpha k_1} \dots \mathring{\tau}_{\alpha k_\chi}, \tag{1.53}$$

$$\partial^\chi u_i \equiv \frac{\partial^\chi u_i}{\partial x_{k_1} \dots \partial x_{k_\chi}}, \qquad \partial^\chi \phi_i \equiv \frac{\partial^\chi \phi_i}{\partial x_{k_1} \dots \partial x_{k_\chi}}. \tag{1.54}$$

Furthermore, conservation of linear momentum, eqn. (1.33), and conservation of moment of momentum, eqn. (1.34), are written as

$$\sum_{\alpha=1}^{n} \sum_{\chi=1}^{M} (-1)^{\chi+1} \frac{(\eta_\alpha)^{\chi-1}}{\chi!} T_{\alpha(\chi)}\, \partial^\chi (p_{\alpha i} + t_{\alpha i}) + F_i = \rho\, \ddot{u}_i, \tag{1.55}$$

$$\sum_{\alpha=1}^{n} \Bigg[\epsilon_{ijq}\, \mathring{\tau}_{\alpha j}\, t_{\alpha q} + \sum_{\chi=1}^{M} (-1)^{\chi+1} \frac{(\eta_\alpha)^{\chi-1}}{\chi!} T_{\alpha(\chi)}\, \partial^\chi \left(m_{\alpha i} - \frac{1}{2} \eta_\alpha\, \epsilon_{ijq}\, \mathring{\tau}_{\alpha j}\, t_{\alpha q} \right) \Bigg] = 0. \tag{1.56}$$

2. Doublet Thermomechanics

K. Mon and M. Ferrari

2.1 Introduction

It may be said that the field of continuum thermomechanics (CT) was established in its contemporary format with the landmark contributions of the late fifties and early sixties (Coleman and Noll 1959, Coleman and Noll 1963, Gurtin 1965, Truesdell and Noll 1965) Disagreement between researchers in CT has not been infrequent (see e.g. Day (1977), Green and Naghdi (1977), Naghdi (1980), Kestin (1990)) and much of it has centered around using the Clausius-Duhem Inequality (CDI) as the embodiment of the Second Law of Thermodynamics and its use to derive constitutive restrictions in the manner proposed by Coleman and Noll (1963). Day (1977), using an example of a rigid, homogeneous, simple heat conductor with memory, has shown that the CDI is consistent with nonunique values of entropy, i.e. the CDI can be satisfied by any of a range of entropy values. Green and Naghdi (1977) have shown that the CDI predicts, for a class of rigid heat conductors in equilibrium, that if heat is added to the medium, temperature must decrease. Green and Naghdi (1977) proposed use of the Clausius Inequality in conjunction with the concept of an entropy balance law as an alternative to the CDI. The Naghdi-Green formalism has been successfully applied to studies of mixtures of interacting continua (Green and Naghdi 1978) and nonlocal elasticity (Green and Naghdi 1978). For elastic constitutive assumptions, the CDI-based approach of Coleman and Noll (1963) yields identical results to the Naghdi-Green formalism (Naghdi 1980).

Generalized continuum mechanical theories have been proposed, with the objective of modeling continua endowed with a material microstructure (Cosserat and Cosserat 1909, Eringen and Suhubi 1964, Mindlin 1964, Green and Rivlin 1964b, Eringen 1966). These are essentially continuum approaches in that they are based on the modeling assumption that all continuum points are endowed with additional kinematic variables that are somehow representative of the microstructure contained in the "differential volume element" centered at the point. Literature studies on the thermodynamics of the generalized continua are not as abundant as those in continuum thermomechanics, and are generally based on the CDI as the mathematical embodiment of the Second Law (Eringen and Suhubi 1964, Eringen 1966, Stojanović 1972). It may be expected that the use of the CDI in this context be subject to the same objections as those mentioned above for conventional continuum mechanics. On this basis, and without entering into the comparative merits of the various approaches to the thermomechanics of media with or without microstructure, in this chapter we employ the procedure of Green and Naghdi (1977) to determine the constitutive restrictions that are imposed on a class of microstructured media by the entropy balance law and the Clausius Inequality. The analysis is limited to the thermoelastic range, where the choice of thermomechanical formulations has been shown to be immaterial, not only for the cited case of continuum thermoelasticity, but also for the case of micropolar elasticity (Ferrari 1985).

In analogy with the methods of Green and Naghdi (1977, 1979), in this chapter we first identify local energy and entropy balance laws, and make constitutive assumptions as to what variables the functions that appear in these balance laws depend on. Substitution of the constitutive assumption into the balance laws allows simplification of these functional dependencies. Next, an analysis is undertaken to determine what further restrictions are implied by consideration of superposed rigid body motions (Green and Naghdi 1979). (Green and Naghdi, 1979). Finally, the results of the analyses are applied to a study of homogeneous linear elastic doublet mechanics.

2.2 Balance Laws

We start from the differential formulation of energy balance

$$\rho r - \mathrm{div}(\mathbf{q}) + P - \rho \dot{E} = 0, \tag{2.1}$$

where the superscripted dot ($\dot{}$) denotes the material time derivative. In eqn. (2.1), r is the volume rate of heat supplied per unit mass, $\mathbf{q}$ is the heat flux vector, P is the mechanical power per unit volume, E is the internal energy per unit mass, and ρ is the mass density.

The differential law of entropy balance, introduced by Green and Naghdi (1977), is

$$\rho\left(\frac{r}{\theta} + \xi\right) - \mathrm{div}\left(\frac{\mathbf{q}}{\theta}\right) = \rho \dot{s}, \tag{2.2}$$

where ξ is the internal rate of entropy production per unit mass, s is the entropy per unit mass, and θ is a function of empirical temperature, T, and other constitutive variables such that $\theta \geq 0$ and $\partial\theta/\partial T \geq 0$. The combination of eqns. (2.1) and (2.2) yields the relation

$$\rho r - \mathrm{div}(\mathbf{q}) = \rho\theta\dot{s} - \rho\theta\xi - \frac{\mathbf{q}\cdot\mathbf{g}}{\theta} = \rho\dot{E} - P, \tag{2.3}$$

where $\mathbf{g}$ denotes grad(θ). Rewriting eqn. (2.3) yields

$$-\rho(\dot{E} - \theta\dot{s}) - \frac{\mathbf{q}\cdot\mathbf{g}}{\theta} - \rho\theta\xi + P = 0, \tag{2.4}$$

or, in terms of the Helmholtz free energy ($\psi = E - \theta s$)

$$-\rho(\dot{\psi} + \dot{\theta} s) - \frac{\mathbf{q} \cdot \mathbf{g}}{\theta} - \rho\theta\xi + P = 0. \tag{2.5}$$

Expressions (2.4) and (2.5) will be referred to as energy/entropy balance equations and must hold for all thermomechanical processes. Definition of a thermomechanical process will be delayed until the mechanical power is discussed further.

The only difference in these equations, between continuum mechanics (CM) and doublet mechanics (DM), lies in how the mechanical power per unit volume P is written. In CM, $P = \boldsymbol{\sigma} \cdot \mathbf{L}$ where $\boldsymbol{\sigma}$ is the stress tensor and $\mathbf{L}$ is the velocity gradient. $\mathbf{L}$ is related to the deformation gradient, $\mathbf{F}$, by $\mathbf{L} = \dot{\mathbf{F}}\mathbf{F}^{-1}$. In DM,

$$P = \sum_{\alpha=1}^{n} (p_\alpha \dot{\epsilon}_\alpha + m_\alpha \dot{\mu}_\alpha + \mathbf{t}_\alpha \cdot \dot{\boldsymbol{\gamma}}_\alpha). \tag{2.6}$$

The elongation microstress p_α is conjugate to the elongation microstrain ϵ_α, the torsional microstress m_α is conjugate to the torsional microstrain μ_α, and the shear microstress vector $\mathbf{t}_\alpha$ is conjugate to the shear microstrain vector, γ_α. Use has been made of the fact that $\mathbf{p}_\alpha \cdot \boldsymbol{\epsilon}_\alpha = p_\alpha \epsilon_\alpha$ and $\mathbf{m}_\alpha \cdot \boldsymbol{\mu}_\alpha = m_\alpha \mu_\alpha$, i.e. the elongation and torsion microstrains and stresses are collinear with the α doublet axis.

Additionally, the microstresses are required to satisfy the balance of linear momentum,

$$\sum_{\alpha=1}^{n} \sum_{\chi=1}^{M} (-1)^{\chi-1} \frac{\eta_\alpha^{\chi-1}}{\chi!} \overset{\circ}{\tau}_{\alpha k_1} \cdots \overset{\circ}{\tau}_{\alpha k_\chi} \frac{\partial^\chi (p_{\alpha i} + t_{\alpha i})}{\partial x_{k_1} \cdots \partial x_{k_\chi}} + F_i = \rho \ddot{u}_i, \tag{2.7}$$

and moment of momentum,

$$\begin{aligned} &\sum_{\alpha=1}^{n} \Bigg(\epsilon_{ijq} \overset{\circ}{\tau}_{\alpha j} t_{\alpha q} \\ &\qquad + \sum_{\chi=1}^{M} (-1)^{\chi-1} \frac{\eta_\alpha^{\chi-1}}{\chi!} \overset{\circ}{\tau}_{\alpha k_1} \cdots \overset{\circ}{\tau}_{\alpha k_\chi} \frac{\partial^\chi (m_{\alpha i} - \frac{1}{2} \eta_\alpha \epsilon_{ijq} \overset{\circ}{\tau}_{\alpha j} t_{\alpha q})}{\partial x_{k_1} \cdots \partial x_{k_\chi}} \Bigg) \\ &\qquad + L_i = 0, \end{aligned} \tag{2.8}$$

where F_i is the volume body force, L_i is the volume distribution of body couples, and ϵ_{ijk} is the permutation tensor. The summing limit, M, refers to the degree of approximation.

The independent variables in the above treatment are

$$\{\mathbf{u}, \boldsymbol{\phi}, T\}. \tag{2.9}$$

The balance laws (2.2), (2.5), (2.7), and (2.8) contain the fields

$$\{\psi, \theta, s, \xi, p_\alpha, m_\alpha, \mathbf{t}_\alpha, \mathbf{q}\}, \tag{2.10}$$

as well as

$$\{\mathbf{F}, \mathbf{L}, r\}. \tag{2.11}$$

We assume that the fields (2.10) depend constitutively on the variables in eqn. (2.9) and possibly on their space and time derivatives. We assume that:

1. The balance laws hold for arbitrary choices of the variables in (2.9) and, if constitutive assumptions require, their space and time derivatives;
2. The fields (2.10) are calculated from their constitutive equations;
3. The fields (2.11) can then be found from the balance of momentum (2.7), moment of momentum (2.8), and entropy (2.2);
4. The energy/entropy balance equation (2.5) may be imposed as an identity for every choice of variables (2.9). This allows restrictions on the constitutive equations to be derived.

A thermomechanical process is defined by specifying the set of variables (2.9) such that the balance laws are satisfied.

2.3 Elastic Constitutive Assumptions

The constitutive assumptions of doublet elasticity are

$$\psi = \hat{\psi}(\epsilon_\alpha, \mu_\alpha, \gamma_\alpha, T, \nabla T : \mathbf{X}), \tag{2.12}$$

$$s = \hat{s}(\epsilon_\alpha, \mu_\alpha, \gamma_\alpha, T, \nabla T : \mathbf{X}), \tag{2.13}$$

$$\xi = \hat{\xi}(\epsilon_\alpha, \mu_\alpha, \gamma_\alpha, T, \nabla T : \mathbf{X}), \tag{2.14}$$

$$\theta = \hat{\theta}(\epsilon_\alpha, \mu_\alpha, \gamma_\alpha, T, \nabla T : \mathbf{X}), \tag{2.15}$$

$$\mathbf{q} = \hat{\mathbf{q}}(\epsilon_\alpha, \mu_\alpha, \gamma_\alpha, T, \nabla T : \mathbf{X}), \tag{2.16}$$

$$p_\alpha = \hat{p}_\alpha(\epsilon_\beta, \mu_\beta, \gamma_\beta, T, \nabla T : \mathbf{X}), \tag{2.17}$$

$$m_\alpha = \hat{m}_\alpha(\epsilon_\beta, \mu_\beta, \gamma_\beta, T, \nabla T : \mathbf{X}), \tag{2.18}$$

$$\mathbf{t}_\alpha = \hat{\mathbf{t}}_\alpha(\epsilon_\beta, \mu_\beta, \gamma_\beta, T, \nabla T : \mathbf{X}), \tag{2.19}$$

where T is the empirical temperature and α and β range from 1 to n. The $\mathbf{X}$ indicates a possible spatial dependence. Expanding these functions into their partial derivatives and substitution, along with eqn. (2.6), into eqn. (2.5) yields an equation of the form

$$A\dot{T} + B_i \frac{\partial T}{\partial x_i} + C_{ij} \frac{\partial^2 T}{\partial x_i \partial x_j} + \sum_{\alpha=1}^{n} \Big(H1_\alpha \dot{\epsilon}_\alpha + H2_\alpha \dot{\mu}_\alpha + H3_{\alpha i} \dot{\gamma}_{\alpha i} + \cdots$$

$$+ I1_{\alpha i} \frac{\partial \epsilon_\alpha}{\partial x_i} + I2_{\alpha i} \frac{\partial \mu_\alpha}{\partial x_i} + I3_{\alpha ij} \frac{\partial \gamma_{\alpha j}}{\partial x_i} \Big) + G = 0, \tag{2.20}$$

with coefficients

$$A = -\rho\left(\frac{\partial\psi}{\partial T} + s\frac{\partial\theta}{\partial T}\right), \tag{2.21}$$

$$B_i = -\rho\left(\frac{\partial\psi}{\partial\left(\partial T/\partial x_j\right)} + s\frac{\partial\theta}{\partial\left(\partial T/\partial x_j\right)}\right), \tag{2.22}$$

$$C_{ij} = -\frac{q_i}{\theta}\frac{\partial\theta}{\partial\left(\partial T/\partial x_j\right)}, \tag{2.23}$$

$$G = -\frac{q_i}{\theta}\left(\frac{\partial\theta}{\partial T}\frac{\partial T}{\partial x_i} + \frac{\partial\theta}{\partial x_i}\right) - \rho\xi\theta, \tag{2.24}$$

$$H1_\alpha = -\rho\left(\frac{\partial\psi}{\partial\epsilon_\alpha} + s\frac{\partial\theta}{\partial\epsilon_\alpha}\right) + p_\alpha, \tag{2.25}$$

$$H2_\alpha = -\rho\left(\frac{\partial\psi}{\partial\mu_\alpha} + s\frac{\partial\theta}{\partial\mu_\alpha}\right) + m_\alpha, \tag{2.26}$$

$$H3_{\alpha i} = -\rho\left(\frac{\partial\psi}{\partial\gamma_{\alpha i}} + s\frac{\partial\theta}{\partial\gamma_{\alpha i}}\right) + t_{\alpha i}, \tag{2.27}$$

$$I1_{\alpha i} = -\frac{q_i}{\theta}\frac{\partial\theta}{\partial\epsilon_\alpha}, \tag{2.28}$$

$$I2_{\alpha i} = -\frac{q_i}{\theta}\frac{\partial\theta}{\partial\mu_\alpha}, \tag{2.29}$$

$$I3_{\alpha ij} = -\frac{q_i}{\theta}\frac{\partial\theta}{\partial\gamma_{\alpha j}}. \tag{2.30}$$

The vectorial subscripts i and j vary from 1 to 3, while the doublet numbering subscript, α, varies from 1 to n. Equation (2.20) must hold for every thermodynamic process, including arbitrary choices of the functions in the set

$$\left\{\dot{T}, \frac{\partial\dot{T}}{\partial x_i}, \frac{\partial^2 T}{\partial x_i\partial x_j}, \dot{\epsilon}_\alpha, \dot{\mu}_\alpha, \dot{\gamma}_{\alpha i}, \frac{\partial\epsilon_\alpha}{\partial x_i}, \frac{\partial\mu_\alpha}{\partial x_i}, \frac{\partial\gamma_{\alpha j}}{\partial x_i}\right\}. \tag{2.31}$$

Since these functions do not enter into constitutive assumptions (2.12)–(2.19) or the coefficients defined in eqns. (2.21)–(2.30), any functional relationships derived from arbitrary choices of the functions in the set (2.31) must hold for every thermodynamic process. In particular,

a) Taking each member of the set (2.31) to be zero yields $G \equiv 0$.
b) Given that $G = 0$, and taking all of set (2.31) except $\partial^2 T/\partial x_i\partial x_j$ to be zero yields $C_{ij} \equiv 0$. This means that $\theta = \hat{\theta}(\epsilon_\alpha, \mu_\alpha, \gamma_\alpha, T : \mathbf{X})$ i.e. θ is not a function of ∇T.
c) Given that $C_{ij} = G = 0$ and taking all of the set (2.31) except $\partial\dot{T}/\partial x_i$ to be zero yields $B_i \equiv 0$. This means ψ is not a function of ∇T.
d) Given that $C_{ij} = G = B_i = 0$ and taking all of the set (2.31) except $\dot{T}$ to be zero yields $A \equiv 0$.
e) Given that $C_{ij} = G = B_i = A = 0$ and taking all of the set (2.31) except $\partial\epsilon_\alpha/\partial x_i$ to be zero yields $I1_{\alpha i} \equiv 0$. This means θ is not a function of ϵ_α.

f) Given that $C_{ij} = G = B_i = A = I1_{\alpha i} = 0$ and taking all of the set (2.31) except $\partial\mu_\alpha/\partial x_i$ to be zero yields $I2_{\alpha i} \equiv 0$. This means θ is not a function of μ_α.
g) Given that $C_{ij} = G = B_i = A = I1_{\alpha i} = I2_{\alpha i} = 0$ and taking all of the set (2.31) except $\partial\gamma_{\alpha j}/\partial x_i$ to be zero yields $I3_{\alpha ij} \equiv 0$. This means θ is not a function of $\gamma_{\alpha i}$. At this point $\theta = \hat{\theta}(T : \mathbf{X})$.
h) Given that $C_{ij} = G = B_i = A = I1_{\alpha i} = I2_{\alpha i} = I3_{\alpha ij} = 0$ and taking all of the set (2.31) except $\dot{\epsilon}_\alpha$ to be zero yields $H1_\alpha \equiv 0$. This means $p_\alpha = \rho\partial\psi/\partial\epsilon_\alpha$ and, since ψ is not a function of ∇T, p_α is not a function of ∇T.
i) Given that $C_{ij} = G = B_i = A = I1_{\alpha i} = I2_{\alpha i} = I3_{\alpha ij} = H1_\alpha = 0$ and taking all of the set (2.31) except $\dot{\mu}_\alpha$ to be zero yields $H2_\alpha \equiv 0$. This means $m_\alpha = \rho\partial\psi/\partial\mu_\alpha$ and m_α is not a function of ∇T.
j) Given that $C_{ij} = G = B_i = A = I1_{\alpha i} = I2_{\alpha i} = I3_{\alpha ij} = H1_\alpha = H2_\alpha = 0$ and taking all of the set (2.31) except $\dot{\gamma}_{\alpha i}$ to be zero yields $H3_{\alpha i} \equiv 0$. This means $t_{\alpha i} = \rho\partial\psi/\partial\gamma_{\alpha i}$ and $t_{\alpha i}$ is not a function of ∇T.

Up to this point we have not applied the Second Law of Thermodynamics; nothing has been said of which processes are possible and which are not. In this chapter, we retain the viewpoint that different statements of the Second Law are possible, each embodying some aspects of it. In what follows we explore the consequences of the Clausius Inequality (Green and Naghdi 1977),

$$\oint_I \left(\int_{\partial B_t} \frac{\mathbf{q}\cdot\mathbf{n}}{\theta} da - \int_{B_t} \frac{\rho r}{\theta} dv \right) dt \geq 0, \tag{2.32}$$

which states that the sum of the entropy from heat conduction $\mathbf{q}$ through the surface of the body ∂B_t and from radiation r in the body B_t, considered over a closed cycle I, must be greater than or equal to zero. A closed cycle is one in which the entropy, s, is the same before and after the cycle is completed. If the Second Law applies locally to every part of the body, through the use of the divergence theorem and entropy balance (2.2), the Clausius Inequality (2.32) can be reduced to

$$\oint_I \xi \cdot dt \geq 0. \tag{2.33}$$

If the entropy production rate, ξ, is independent of time, eqn. (2.33) implies $\xi \geq 0$.

To incorporate eqn. (2.33) into the above analysis, one must reconsider the results of part a) of the application of the constitutive assumptions (2.12)–(2.19) to the energy/entropy balance equation (2.5), particularly

$$G = -\frac{q_i}{\theta}\left(\frac{\partial\theta}{\partial T}\frac{\partial T}{\partial x_i} + \frac{\partial\theta}{\partial x_i}\right) - \rho\xi\theta = 0. \tag{2.34}$$

Since it is possible to choose the cycle (I) in eqn. (2.33) such that $\mathbf{q}$, T, ∇T, and therefore ξ are independent of time, it must be that $\xi \geq 0$ for all processes. Recalling that $\theta \geq 0$, equation (2.33) is then equivalent to the condition

$$-q_i \left(\frac{\partial \theta}{\partial T} \frac{\partial T}{\partial x_i} + \frac{\partial \theta}{\partial x_i} \right) \geq 0. \tag{2.35}$$

If we assume Fourier's Law of heat conduction holds, i.e. $q_i = K_{ij}\, \partial T / \partial x_j$ when $\partial T / \partial x_j$ approaches zero, with

$$\mathbf{K} = \hat{\mathbf{K}}(\epsilon_\alpha, \mu_\alpha, \gamma_\alpha, T : \mathbf{X}), \tag{2.36}$$

eqn. (2.35) can be rewritten as

$$K_{ij} \frac{\partial \theta}{\partial T} \frac{\partial T}{\partial x_i} \frac{\partial T}{\partial x_j} + K_{ij} \frac{\partial \theta}{\partial x_i} \frac{\partial T}{\partial x_j} \leq 0. \tag{2.37}$$

Given that $\mathbf{K} \neq \mathbf{0}$, if eqn. (2.37) is to hold for arbitrary (near zero) values of $\partial T / \partial x_j$ it must be that $\partial \theta / \partial x_i = 0$. Thus, θ is not an explicit function of $\mathbf{X}$. This result, added to the results of parts d) to f) above, show that θ is an explicit function of T only. Given that $\theta \geq 0$ and $\partial \theta / \partial T \geq 0$, $\hat{\theta}(T)$ is an invertible function and can replace T as a constitutive variable in all of the foregoing equations and relations.

Armed with this result, conclusion d) above is reinterpreted to be

$$\frac{\partial \psi}{\partial \theta} = -s, \tag{2.38}$$

and s is not a function of $\mathbf{g} = \mathrm{grad}(\theta)$ because ψ is not from part c) above. Also (2.34) can be rewritten as

$$\rho \xi \theta = -\frac{q_i}{\theta} g_i, \tag{2.39}$$

and, from eqn. (2.3),

$$\rho r - \mathrm{div}\mathbf{q} = \rho \theta \dot{s}. \tag{2.40}$$

Relations (2.39) and (2.40) are consistent with the results of Coleman and Noll (1963) and Green and Naghdi (1977).

At this point the relations (2.12)–(2.19) can be rewritten as

$$\psi = \hat{\psi}(\epsilon_\alpha, \mu_\alpha, \gamma_\alpha, \theta : \mathbf{X}), \tag{2.41}$$
$$s = \hat{s}(\epsilon_\alpha, \mu_\alpha, \gamma_\alpha, \theta : \mathbf{X}), \tag{2.42}$$
$$\xi = \hat{\xi}(\epsilon_\alpha, \mu_\alpha, \gamma_\alpha, \theta, \mathbf{g} : \mathbf{X}), \tag{2.43}$$
$$\theta = \hat{\theta}(T), \tag{2.44}$$
$$\mathbf{q} = \hat{\mathbf{q}}(\epsilon_\alpha, \mu_\alpha, \gamma_\alpha, \theta, \mathbf{g} : \mathbf{X}), \tag{2.45}$$
$$p_\alpha = \hat{p}_\alpha(\epsilon_\beta, \mu_\beta, \gamma_\beta, \theta : \mathbf{X}), \tag{2.46}$$
$$m_\alpha = \hat{m}_\alpha(\epsilon_\beta, \mu_\beta, \gamma_\beta, \theta : \mathbf{X}), \tag{2.47}$$
$$\mathbf{t}_\alpha = \hat{\mathbf{t}}_\alpha(\epsilon_\beta, \mu_\beta, \gamma_\beta, \theta : \mathbf{X}). \tag{2.48}$$

2.4 Superposed Rigid Body Motions

In the various theories of solid mechanics, once a set of measures of the deformation are chosen, these measures are tested against transformations of the deformed configuration to check whether they are properly invariant. Thus, in finite deformation theories of classical continuum mechanics, the Cauchy-Green tensor $\mathbf{C} = \mathbf{F}^T\mathbf{F}$ is chosen as a strain measure and is determined to be invariant under arbitrary finite rigid motion of the deformed configuration. In geometrically linear continuum theories, $\varepsilon = \text{sym}(\text{grad}(\mathbf{u}))$ is chosen, and is proven to be invariant only under infinitesimal rotations of the deformed configuration. As shown by Casey and Naghdi (1981), invariance of ε under arbitrary finite rotations may be proven, upon choosing to remove the translation and rotation at a "pivot point" in the body from the description of the deformation. This introduces an element of frame specificity to the theory, but allows the linear theory to be deduced as a special subcase of the finite theory.

The set of deformations appropriate to a theory is chosen on the basis of the ability of said measures to quantify those aspects of the deformation that are of interest within the theory itself. Thus, $\tilde{\epsilon}$ is an appropriate measure in linear continuum theories in that it contains the desired information on changes in lengths, areas, volumes, and angles. By comparison with this, the lack of proper invariance of $\tilde{\epsilon}$ under finite rotations has been considered a notion of lesser importance throughout the history of mechanics. As will be shown, the kinematic equations of DM are properly invariant under superposed rigid body motions, an undervalued but highly desirable component of any mechanical theory.

In the present form of doublet mechanics, as presented in the original paper by Granik and Ferrari (1993) and summarized in Chap. 1, the measures of deformation are the elongational strains ϵ_α, the torsional strains μ_α, and the shear strains γ_α. In this and the following sections we consider the geometrically linear kinematic relations

$$\epsilon_\alpha = \frac{\boldsymbol{\tau}_\alpha^\circ \cdot \Delta\mathbf{u}_\alpha}{\eta_\alpha}, \tag{2.49}$$

$$\mu_\alpha = \frac{\boldsymbol{\tau}_\alpha^\circ \cdot \Delta\boldsymbol{\phi}_\alpha}{\eta_\alpha}, \tag{2.50}$$

$$\gamma_\alpha = -\left(\boldsymbol{\phi} + \frac{1}{2}\Delta\boldsymbol{\phi}_\alpha - \boldsymbol{\tau}_\alpha^\circ \times \boldsymbol{\tau}_\alpha\right) \times \boldsymbol{\tau}_\alpha^\circ, \tag{2.51}$$

where η_α is the internodal distance of the α-th particle doublet, $\boldsymbol{\tau}_\alpha^\circ$ is a unit vector in the undeformed configuration oriented along the α-th doublet axis, and $\Delta\mathbf{u}_\alpha$ and $\Delta\boldsymbol{\phi}_\alpha$ are given by the following relations in Cartesian coordinates:

$$\left.\begin{array}{l}\Delta u_{\alpha i}(\mathbf{X},t)\\ \Delta\phi_{\alpha i}(\mathbf{X},t)\end{array}\right\} = \sum_{\chi=1}^{M} \frac{\eta_\alpha^\chi}{\chi!}\tau_{\alpha k_1}^\circ \cdots \tau_{\alpha k_\chi}^\circ \frac{\partial^\chi}{\partial x_{k_1}\cdots\partial x_{k_\chi}}\left\{\begin{array}{l}u_i(\mathbf{X},t)\\ \phi_i(\mathbf{X},t)\end{array}\right|_{\mathbf{X}=\mathbf{X}_\circ}. \tag{2.52}$$

Repeated Latin indices (e.g., k_j) should be summed over while repeated Greek indices (e.g., α) should not. This equation should be evaluated only at the doublet nodes (when $\mathbf{X} = \mathbf{X}_\circ$). The vector fields of translations $\mathbf{u}$ and rotations $\boldsymbol{\phi}$ are mutually independent and are functions of position $\mathbf{X}$ and time t.

In this section the invariance properties of these measures of deformation are investigated under the transformations

$$u_i \rightarrow u_i^+ \equiv Q_{ij}(t)u_j + c_i(t), \tag{2.53}$$

$$\phi_i \rightarrow \phi_i^+ \equiv Q_{ij}(t)\phi_j, \tag{2.54}$$

where $Q_{ij}(t)$ represents a proper orthogonal matrix (a rotation) and $c_i(t)$ represents a rigid translation. Under eqns. (2.53) and (2.54), the doublet unit vectors transform as

$$\overset{\circ}{\tau}{}^{+}_{\alpha i} = Q_{ij}\overset{\circ}{\tau}_{\alpha i}, \tag{2.55}$$

and the internodal distances remain unaltered, i.e., $\eta_\alpha^+ = \eta_\alpha$.

2.4.1 Microstrains

Using relations (2.53)–(2.55) and the identities

$$Q_{ij}Q_{ik} = \delta_{jk} \quad \text{and} \quad \frac{\partial v_i^+}{\partial x_j^+} = Q_{im}Q_{jk}\frac{\partial v_m}{\partial x_k}, \tag{2.56}$$

where $\mathbf{v}$ is any vector, it can be deduced from eqn. (2.52) that

$$\Delta u_{\alpha i}^+ = Q_{ij}\Delta u_{\alpha j} \quad \text{and} \quad \Delta\phi_{\alpha i}^+ = Q_{ij}\Delta\phi_{\alpha j}, \tag{2.57}$$

and from eqns. (2.49), (2.50) and (2.53)–(2.57) that

$$\epsilon_\alpha^+ = \epsilon_\alpha \quad \text{and} \quad \mu_\alpha^+ = \mu_\alpha. \tag{2.58}$$

Relation (2.51) for the microshear strain vector can be rewritten in Cartesian form as

$$\gamma_{\alpha i} = -\left[(\phi_j + \frac{1}{2}\Delta\phi_{\alpha j})\overset{\circ}{\tau}_{\alpha p}\epsilon_{ijp} + \left(\frac{\overset{\circ}{\tau}_{\alpha i}\overset{\circ}{\tau}_{\alpha j} - \delta_{ij}}{\eta_\alpha}\right)\Delta u_{\alpha j}\right], \tag{2.59}$$

where ϵ_{ijk} is the permutation tensor. Again using the relations (2.53)–(2.57) it is found that

$$\gamma_{\alpha i}^+ = -\left[(\phi_l + \frac{1}{2}\Delta\phi_{\alpha l})Q_{jl}Q_{pq}\overset{\circ}{\tau}_{\alpha q}\epsilon_{ijp} + \left(\frac{Q_{il}\overset{\circ}{\tau}_{\alpha l}\overset{\circ}{\tau}_{\alpha q} - Q_{iq}}{\eta_\alpha}\right)\Delta u_{\alpha q}\right], \tag{2.60}$$

with no further algebraic simplification apparent. We note from eqns. (2.59) and (2.60) that $\gamma_\alpha^+ = \mathbf{Q}\gamma_\alpha$ if $\mathbf{Q}$ is such that

$$Q_{jm}Q_{il}Q_{pq}\epsilon_{jip} = \epsilon_{mlq}. \tag{2.61}$$

Equation (2.61) is an identity for all $\mathbf{Q}$ given that the determinant of $\mathbf{Q}$ is unitary (Shames and Cozzarelli 1991).

By eqns. (2.58) and (2.61) it is seen that the axial, torsional, and shear microstrains are properly invariant under the finite rigid motions considered in eqns. (2.53) and (2.54).

2.4.2 Microstresses

It is postulated that the doublet mechanical microstresses are invariant, apart from changes in orientation, under the transformations (2.53) and (2.54)

$$p_\alpha^+ = p_\alpha, \qquad m_\alpha^+ = m_\alpha, \qquad t_{\alpha i}^+ = Q_{ij} t_{\alpha i}. \tag{2.62}$$

Under the constitutive assumptions (2.17)–(2.19), relations (2.62) imply that

$$\hat{p}_\alpha(\epsilon_\beta, \mu_\beta, \mathbf{Q}\boldsymbol{\gamma}_\beta, \theta : \mathbf{X}) = \hat{p}_\alpha(\epsilon_\beta, \mu_\beta, \boldsymbol{\gamma}_\beta, \theta : \mathbf{X}), \tag{2.63}$$

$$\hat{m}_\alpha(\epsilon_\beta, \mu_\beta, \mathbf{Q}\boldsymbol{\gamma}_\beta, \theta : \mathbf{X}) = \hat{m}_\alpha(\epsilon_\beta, \mu_\beta, \boldsymbol{\gamma}_\beta, \theta : \mathbf{X}), \tag{2.64}$$

$$\hat{\mathbf{t}}_\alpha(\epsilon_\beta, \mu_\beta, \mathbf{Q}\boldsymbol{\gamma}_\beta, \theta : \mathbf{X}) = \mathbf{Q}\hat{\mathbf{t}}_\alpha(\epsilon_\beta, \mu_\beta, \boldsymbol{\gamma}_\beta, \theta : \mathbf{X}). \tag{2.65}$$

Relations (2.63)–(2.65) place significant restrictions on the functional forms of the microstresses. Through application of Cauchy's representation theorems (Truesdell and Noll 1965) eqns. (2.63)–(2.65) can be rewritten as

$$p_\alpha = \hat{p}_\alpha(\epsilon_\beta, \mu_\beta, \gamma_\beta, \theta : \mathbf{X}), \tag{2.66}$$

$$m_\alpha = \hat{m}_\alpha(\epsilon_\beta, \mu_\beta, \gamma_\beta, \theta : \mathbf{X}), \tag{2.67}$$

$$\mathbf{t}_\alpha = \sum_{\beta=1}^{n} \hat{\mathbf{I}}_{\alpha\beta}(\epsilon_\beta, \mu_\beta, \gamma_\beta, \theta : \mathbf{X}) \boldsymbol{\gamma}_\beta, \tag{2.68}$$

where γ_β is the magnitude of $\boldsymbol{\gamma}_\beta$.

2.4.3 Other Functions

Similar restrictions for the other functions in our development are obtained in the same manner as above

$$\psi = \hat{\psi}(\epsilon_\alpha, \mu_\alpha, \gamma_\alpha, \theta : \mathbf{X}), \tag{2.69}$$

$$s = \hat{s}(\epsilon_\alpha, \mu_\alpha, \gamma_\alpha, \theta : \mathbf{X}), \tag{2.70}$$

$$\xi = \hat{\xi}(\epsilon_\alpha, \mu_\alpha, \gamma_\alpha, \theta : \mathbf{X}), \tag{2.71}$$

$$\mathbf{q} = \sum_{\alpha=1}^{n} \hat{\mathbf{H}}_\alpha(\epsilon_\beta, \mu_\beta, \gamma_\beta, \theta, g, \boldsymbol{\gamma}_\beta \cdot \mathbf{g} : \mathbf{X}) \boldsymbol{\gamma}_\alpha + \cdots + \hat{\mathbf{K}}(\epsilon_\beta, \mu_\beta, \gamma_\beta, \theta, g, \boldsymbol{\gamma}_\beta \cdot \mathbf{g} : \mathbf{X}) \mathbf{g}, \tag{2.72}$$

where g is the magnitude of grad(θ). The heat flux vector $\mathbf{q}$ in (2.72) is shown to depend not only on the temperature gradient $\mathbf{g}$, as in CM, but also on the shear microstrain $\boldsymbol{\gamma}_\alpha$. In DM, $\mathbf{q}$ depends constitutively (see eqn. (2.45)) on these two vectors and not just one vector $\mathbf{g}$ as in CM. As will be shown, the dependence of $\mathbf{q}$ on $\boldsymbol{\gamma}_\alpha$ can be removed in the linear theory.

2.5 Linear Elastic Doublet Mechanics

Assuming material homogeneity, the most general physically linear relation between the microstrains and microstresses takes the form

$$p_\alpha = \sum_{\beta=1}^{n}(A_{\alpha\beta}\,\epsilon_\beta + B_{\alpha\beta}\,\mu_\beta + C_{\alpha\beta}\,\gamma_\beta) + J_\alpha\Theta, \tag{2.73}$$

$$m_\alpha = \sum_{\beta=1}^{n}(D_{\alpha\beta}\,\epsilon_\beta + E_{\alpha\beta}\,\mu_\beta + F_{\alpha\beta}\,\gamma_\beta) + K_\alpha\Theta, \tag{2.74}$$

$$t_{\alpha i} = \sum_{\beta=1}^{n} I_{\alpha\beta ij}\,\gamma_{\beta j}, \tag{2.75}$$

where $\Theta \equiv \theta - \theta_\circ$ is an increment of the temperature and $\theta_\circ$ is the temperature of the granular media in an initial state.

Further restrictions on the micromoduli follow from the results of parts h) through j) of the application of the constitutive assumption to the energy/entropy balance equation and the mathematical requirement that partial differentiation be independent of the sequence of differentiation, i.e.,

$$\rho\frac{\partial^2\psi}{\partial\epsilon_\alpha\partial\epsilon_\beta} = \rho\frac{\partial^2\psi}{\partial\epsilon_\beta\partial\epsilon_\alpha} = \frac{\partial p_\alpha}{\partial\epsilon_\beta} = \frac{\partial p_\beta}{\partial\epsilon_\alpha} = A_{\alpha\beta} = A_{\beta\alpha}. \tag{2.76}$$

Such considerations lead to the conclusions that

$$\left.\begin{array}{lll} A_{\alpha\beta} = A_{\beta\alpha} & E_{\alpha\beta} = E_{\beta\alpha} & I_{\alpha\beta ij} = I_{\beta\alpha ji} \\ B_{\alpha\beta} = D_{\beta\alpha} & C_{\alpha\beta} = F_{\alpha\beta} = 0. & \end{array}\right\} \tag{2.77}$$

Using relations (2.77), relations (2.73)–(2.75) are rewritten as

$$p_\alpha = \sum_{\beta=1}^{n}(A_{\alpha\beta}\,\epsilon_\beta + B_{\alpha\beta}\,\mu_\beta) + J_\alpha\Theta, \tag{2.78}$$

$$m_\alpha = \sum_{\beta=1}^{n}(D_{\alpha\beta}\,\epsilon_\beta + E_{\alpha\beta}\,\mu_\beta) + K_\alpha\Theta, \tag{2.79}$$

$$t_{\alpha i} = \sum_{\beta=1}^{n} I_{\alpha\beta ij}\,\gamma_{\beta j}, \tag{2.80}$$

with the conditions $A_{\alpha\beta} = A_{\beta\alpha}$, $E_{\alpha\beta} = E_{\beta\alpha}$, $I_{\alpha\beta ij} = I_{\beta\alpha ji}$, and $B_{\alpha\beta} = D_{\beta\alpha}$.

In the linear regime, we may rewrite the expression for the heat flux vector (2.72) as

$$\mathbf{q} = \sum_{\alpha=1}^{n}\mathbf{H}_\alpha\gamma_\alpha + \mathbf{K}\,\mathbf{g}, \tag{2.81}$$

where $\mathbf{H}_\alpha$ and $\mathbf{K}$ are no longer functions of the constitutive variables as in the nonlinear regime (see eqn. (2.72)). Due to the kinematic nature of the microshear, i.e. the dependence of γ_α on the doublet geometry, it must be that the heat flux, $\mathbf{q}$, has no dependence on γ_α. To see this, consider a doublet that is perpendicular to the plane of the page (i.e., one node above and one node below). Suppose that, looked at from above, there exists a nonzero microshear strain oriented along the positive horizontal ("x") axis. Looked at from below, the same microshear strain would be oriented along the negative horizontal axis. If the heat flux were to depend on the microshear, one would determine different values for the heat flux vector depending on one's position with respect to the doublet arrangement. Thus,

$$\mathbf{q} = \mathbf{K}\,\mathbf{g}, \tag{2.82}$$

for linear DM as well as for linear CM.

2.6 Closure

From the present thermomechanical analysis it is concluded that:

1. The constitutive functional measure of temperature θ must be a function only of empirical temperature T.
2. The Helmholtz free energy, ψ, is not a function of the temperature gradient, $\mathbf{g} = \mathrm{grad}(\theta)$.
3. The microstresses and microstrains obey the following relations:

$$p_\alpha = \rho\frac{\partial\psi}{\partial\epsilon_\alpha}, \qquad m_\alpha = \rho\frac{\partial\psi}{\partial\mu_\alpha}, \qquad t_{\alpha i} = \rho\frac{\partial\psi}{\partial\gamma_{\alpha i}}, \tag{2.83}$$

   and thus the microstresses are also independent of grad(θ).
4. The Helmholtz free energy and the entropy s are related by:

$$s = -\frac{\partial\psi}{\partial\theta}, \tag{2.84}$$

   and thus the entropy is also independent of grad(θ).
5. The microstrains ϵ_α, μ_α, and γ_α are properly invariant under arbitrary finite rotations

$$\epsilon_\alpha^+ = \epsilon_\alpha, \qquad \mu_\alpha^+ = \mu_\alpha, \qquad \gamma_\alpha^+ = \mathbf{Q}\gamma_\alpha. \tag{2.85}$$

6. For an elastic solid, the functional representations of the microstresses must obey the following relations:

$$p_\alpha = \hat{p}_\alpha(\epsilon_\beta, \mu_\beta, \gamma_\beta, \theta : \mathbf{X}), \tag{2.86}$$

$$m_\alpha = \hat{m}_\alpha(\epsilon_\beta, \mu_\beta, \gamma_\beta, \theta : \mathbf{X}), \tag{2.87}$$

$$\mathbf{t}_\alpha = \sum_{\beta=1}^{n} \hat{\mathbf{I}}_{\alpha\beta}(\epsilon_\beta, \mu_\beta, \gamma_\beta, \theta : \mathbf{X})\gamma_\beta. \tag{2.88}$$

7. The constitutive laws for a homogeneous linear elastic solid reduce to

$$p_\alpha = \sum_{\beta=1}^{n} (A_{\alpha\beta}\,\epsilon_\beta + B_{\alpha\beta}\,\mu_\beta) + J_\alpha\Theta, \tag{2.89}$$

$$m_\alpha = \sum_{\beta=1}^{n} (D_{\alpha\beta}\,\epsilon_\beta + E_{\alpha\beta}\,\mu_\beta) + K_\alpha\Theta, \tag{2.90}$$

$$t_{\alpha i} = \sum_{\beta=1}^{n} I_{\alpha\beta ij}\,\gamma_{\beta j}, \tag{2.91}$$

with the conditions $A_{\alpha\beta} = A_{\beta\alpha}$, $E_{\alpha\beta} = E_{\beta\alpha}$, $I_{\alpha\beta ij} = I_{\beta\alpha ji}$, and $B_{\alpha\beta} = D_{\beta\alpha}$.

3. Multi-component Constitutive Equations

K. Mon and M. Ferrari

3.1 Introduction

The theory of mixtures has been the subject of intensive study in contemporary mechanics (Truesdell 1957, Truesdell 1969, Truesdell and Toupin 1960, Green and Naghdi 1965, Green and Naghdi 1978, Goodman and Cowin 1972, Passman 1977, Nunziato and Walsh 1980) and has spawned as many controversies about the statement and application of the Second Law of Thermodynamics as studies of single component materials. Most mixture theories assume that each component of the mixture occupies space simultaneously and is required to satisfy its own set of balance laws involving additional "growth" terms that allow the components to exchange mass, momentum, and energy among themselves in such a way that these quantities are conserved for the mixture as a whole. Generally, such treatments make use of different temperatures for each component. Controversies arise over which entropy inequality the components are required to satisfy and whether entropy inequalities that apply to the mixture as a whole can be validly applied to each component individually.

Perhaps the most notable generalized continuum mechanical theory of mixtures is that due to Goodman and Cowin (1972) with later improvements by Passman (1977) and extension to reacting mixtures by Nunziato and Walsh (1980). In this theory, the material is considered continuous and each point of the continua is assigned a scalar-valued "director" corresponding to the proportion of the volume that actually contains material (i.e., $(1 - \text{porosity})$). Again we note that the concept of material microstructure is inherent in the doublet mechanical material model obviating the need for generalized continuum techniques.

The treatment presented here is much simplified relative to those mentioned above and represents the first treatment of multicomponent doublet mechanics (DM). Our approach follows classical non-equilibrium treatments, such as those of DeGroot and Mazur (1962) and Kirkaldy and Young (1987), in which each component supports its own body forces and emits its own radiation, however, there is only one set of balance laws for the mixture as a whole. The presented approach differs from the classical treatments mentioned above in that constitutive restrictions are derived in the manner proposed by Green and Naghdi (1977) as applied successfully to single component DM in the previous chapter. We also require at the outset that each component have the same temperature θ. In contrast to the previous treatment of single component materials in which the Helmholtz free energy ψ was used, here we elect to use internal energy e as the thermodynamic potential with which to analyze constitutive restrictions. It is the authors' view that what this treatment lacks in generality is somewhat offset by its ease of applicability and its use of the local kinematical relations of doublet mechanics.

In Sect. 3.2, we consider the concept of mass balance within the framework of a multicomponent theory. In Sect. 3.3, we present the non-linear dynamic equations of motion, using the concept of the barycentric time derivative introduced in Sect. 3.2. In Sect. 3.4, we present the energy/entropy balance equations including a new term representing the chemical work and discuss the subtleties of its application in contemporary thermodynamics. In Sects. 3.6 and 3.7, we briefly present the results of application of the thermomechanical framework to a study of multicomponent elasticity including a more detailed consideration of the process of linearization of the balance laws. Section 3.8 enumerates the constitutive relations of multicomponent linear doublet elasticity. Overall, the presentation is similar to that of the previous chapter with the exception of our use of internal energy as the governing potential, the inclusion of chemical work terms, and the exceptions noted above.

3.2 Mass Balance

We start by analyzing the concept of mass balance within the context of DM. Let us consider a system consisting of m components. Following continuum treatments (DeGroot and Mazur 1962, Kirkaldy and Young 1987), for each component a, the differential law of mass balance is

$$\frac{\partial \rho_a}{\partial t} = -\mathrm{div}(\rho_a \mathbf{v}_a), \tag{3.1}$$

where $\mathbf{v}_a$ is the velocity of component a, and ρ_a is the density of component a. Summing eqn. (3.1) over all components a, we have

$$\frac{\partial \rho}{\partial t} = -\mathrm{div}(\rho \mathbf{v}), \tag{3.2}$$

where $\rho = \sum_{a=1}^{m} \rho_a$ is the total density and $\mathbf{v} = (1/\rho)\sum_{a=1}^{m} \rho_a \mathbf{v}_a$ is the barycentric or center of mass velocity. With the introduction of barycentric time derivative represented by a superposed dot:

$$\dot{\varphi} \equiv \frac{\partial \varphi}{\partial t} + \mathbf{v} \cdot \mathrm{grad}\varphi, \tag{3.3}$$

and the barycentric diffusion flux

$$\mathbf{J}_a = \rho_a(\mathbf{v}_a - \mathbf{v}), \tag{3.4}$$

such that

$$\sum_{a=1}^{m} \mathbf{J}_a = 0, \tag{3.5}$$

eqn. (3.1) can be rewritten

$$\dot{\rho}_a = -\rho_a \operatorname{div}(\mathbf{v}) - \operatorname{div}(\mathbf{J}_a), \tag{3.6}$$

eqn. (3.2) is rewritten as

$$\dot{\rho} = -\rho \operatorname{div}(\mathbf{v}), \tag{3.7}$$

or, using mass fractions $c_a = \rho_a/\rho$ such that

$$\sum_{a=1}^{m} c_a = 1, \tag{3.8}$$

we have

$$\rho \dot{c}_a = -\operatorname{div}(\mathbf{J}_a). \tag{3.9}$$

3.3 Dynamic Equations of Motion

The basic approach taken here is to assume that the material making up each doublet node is composed of possibly m different components each occupying the entire node volume simultaneously. The DM material model dictates that all of the properties ascribed to a node are considered to act at the geometric center of the node, where the doublets intersect. This formulation yields the balance of momentum

$$\sum_{\alpha=1}^{n} \sum_{\chi=1}^{M} (-1)^{\chi-1} \frac{\eta_\alpha^{\chi-1}}{\chi!} \overset{\circ}{\tau}_{\alpha k_1} \cdots \overset{\circ}{\tau}_{\alpha k_\chi} \frac{\partial^\chi (p_{\alpha i} + t_{\alpha i})}{\partial x_{k_1} \cdots \partial x_{k_\chi}} + \sum_{a=1}^{m} \rho_a F_{ai} = \rho \dot{v}_i, \tag{3.10}$$

and moment of momentum

$$\sum_{\alpha=1}^{n} \Bigg(\epsilon_{ijq} \overset{\circ}{\tau}_{\alpha j} t_{\alpha q} + \sum_{\chi=1}^{M} (-1)^{\chi-1} \frac{\eta_\alpha^{\chi-1}}{\chi!} \overset{\circ}{\tau}_{\alpha k_1} \cdots \overset{\circ}{\tau}_{\alpha k_\chi} \frac{\partial^\chi (m_{\alpha i} - \frac{1}{2}\eta_\alpha \epsilon_{ijq} \overset{\circ}{\tau}_{\alpha j} t_{\alpha q})}{\partial x_{k_1} \cdots \partial x_{k_\chi}} \Bigg) + \sum_{a=1}^{m} \rho_a L_{ai} = 0, \tag{3.11}$$

where $\mathbf{F}_a$ is the body force per unit mass acting on component a, $\mathbf{L}_a$ is the distribution of body couples per unit mass on component a, and ϵ_{ijk} is the permutation tensor. v_i is a Cartesian component of the center of mass velocity.

3.4 Energy Balance

The differential formulation of energy balance for the material is

$$r - \mathrm{div}(\mathbf{q}) + P + C - \dot{e} = 0. \tag{3.12}$$

In eqn. (3.12) r is the volume rate of heat supplied per unit volume, $\mathbf{q}$ is the heat flux vector, P and C are the mechanical and chemical power per unit volume, respectively, and e is the internal energy per unit volume. Of course, one could define r as a summation of heat supplies associated with each component, however, in this work, we elect to use the simpler definition given above. This choice leads to no loss of generality since r is determined by the balance laws. One should note the change in the definition of some of the quantities in eqn. (3.12) from those used in Chap. 2. In most of the DM developments thus far, the volume of the DM "unit cell" (defined by the doublets) is implicitly assumed to remain constant from node to node, i.e. there exists inversion symmetry of the doublets defined at each node. Thus, in a DM context, it is highly desirable to define properties in terms of volume densities, particularly for studies involving compositional changes.

The differential law of entropy balance, introduced by Green and Naghdi (1977), is

$$\left(\frac{r}{\theta} + \xi\right) - \mathrm{div}\left(\frac{\mathbf{q}}{\theta}\right) = \dot{s}, \tag{3.13}$$

where ξ is the internal rate of entropy production per unit volume, s is the entropy per unit volume, and θ is a constitutive measure of temperature such that $\theta \geq 0$. The combination of eqns. (3.12) and (3.13) yields the relation

$$r - \mathrm{div}(\mathbf{q}) = \theta\dot{s} - \theta\xi - \frac{\mathbf{q}\cdot\mathbf{g}}{\theta} = \dot{e} - P - C, \tag{3.14}$$

where $\mathbf{g} = \mathrm{grad}(\theta)$. Rewriting eqn. (3.14) yields

$$-\left(\dot{e} - \theta\dot{s}\right) - \frac{\mathbf{q}\cdot\mathbf{g}}{\theta} - \theta\xi + P + C = 0, \tag{3.15}$$

or, in terms of the Helmholtz free energy ($\psi = e - \theta s$)

$$-\left(\dot{\psi} + \dot{\theta}s\right) - \frac{\mathbf{q}\cdot\mathbf{g}}{\theta} - \theta\xi + P + C = 0. \tag{3.16}$$

The mechanical power per unit volume P in DM is

$$P = \sum_{\alpha=1}^{n} (p_\alpha \dot{\epsilon}_\alpha + m_\alpha \dot{\mu}_\alpha + \mathbf{t}_\alpha \cdot \dot{\boldsymbol{\gamma}}_\alpha). \tag{3.17}$$

p_α is the elongation microstress which is conjugate to the elongation microstrain ϵ_α, m_α is the torsional microstress which is conjugate to the torsional microstrain μ_α, and $\mathbf{t}_\alpha$ is the shear microstress vector which is conjugate to the shear microstrain vector, $\boldsymbol{\gamma}_\alpha$.

The chemical power per unit volume C is given by

$$C = \sum_{a=1}^{m} M_a \dot{c}_a, \tag{3.18}$$

where there are m chemical components associated with each doublet node. M_a is interpreted as the chemical potential and c_a as the mass fraction of chemical component a. The chemical power and chemical potential are somewhat subtle concepts in that all permeable system boundaries (walls) are also diathermal; chemical species cannot be added or subtracted from a system without the possibility of a simultaneous transfer of heat. This leads to the ability to consider chemical changes as either changes in the heat supplied to (lost from) or the work done on (by) the system under consideration. Treating chemical changes as a heat term is common in texts on nonequilibrium thermodynamics (DeGroot and Mazur 1962, Kirkaldy and Young 1987) and has the advantage that heat is easily measured since the energy change can be determined from the initial and terminal states of the system and the mechanical work can be determined independently from consideration of the laws of mechanics. The major disadvantage of treating chemical changes as a heat term stems from the loss of the usual association between heat and entropy. Using the "chemical heat" approach,

$$\delta s = \frac{1}{\theta}\left(\delta q - \sum_{a=1}^{m} M_a \delta c_a\right), \tag{3.19}$$

meaning that entropy is not simply the heat divided by temperature. In this work, we elect to consider chemical contributions to the energy change as a chemical work analogous to the familiar concept of mechanical work. Much as the microstresses are the conjugate forces to the microstrains, the chemical potentials are the conjugate forces to changes in the mole fractions. This approach allows the traditional relation between heat and entropy to be maintained.

The independent variables in the above treatment are

$$\{\mathbf{u}, \boldsymbol{\phi}, s, \{c_a\}\}. \tag{3.20}$$

The balance laws (3.13), (3.12), (3.10), and (3.11) contain the fields

$$\{e, \theta, \xi, p_\alpha, m_\alpha, \mathbf{t}_\alpha, \mathbf{q}\}, \tag{3.21}$$

as well as

$$\{\mathbf{F}_a, \mathbf{L}_a, r\}. \tag{3.22}$$

We assume that:

1. The balance laws hold for arbitrary choices of the variables in eqn. (3.20) and, if constitutive assumptions require, their space and time derivatives;
2. The fields (3.21) are calculated from their constitutive equations;
3. The fields (3.22) can then be found from the balance of momentum (3.10), moment of momentum (3.11), and entropy (3.12);

4. The energy/entropy balance equation (3.15) may be imposed as an identity for every choice of variables (3.20). This allows restrictions on the constitutive equations to be derived.

A thermomechanical process is defined by specifying the set of variables (3.20) such that the balance laws are satisfied.

3.5 Elastic Constitutive Assumptions

The constitutive assumptions of doublet elasticity on the material response functions are:

$$e = \hat{e}(\epsilon_\alpha, \mu_\alpha, \gamma_\alpha, s, \mathbf{g}, c_a : \mathbf{X}), \tag{3.23}$$

$$\theta = \hat{\theta}(\epsilon_\alpha, \mu_\alpha, \gamma_\alpha, s, \mathbf{g}, c_a : \mathbf{X}), \tag{3.24}$$

$$\xi = \hat{\xi}(\epsilon_\alpha, \mu_\alpha, \gamma_\alpha, s, \mathbf{g}, c_a : \mathbf{X}), \tag{3.25}$$

$$\mathbf{q} = \hat{\mathbf{q}}(\epsilon_\alpha, \mu_\alpha, \gamma_\alpha, : s, \mathbf{g}, c_a : \mathbf{X}), \tag{3.26}$$

$$p_\alpha = \hat{p}_\alpha(\epsilon_\beta, \mu_\beta, \gamma_\beta, s, \mathbf{g}, c_a : \mathbf{X}), \tag{3.27}$$

$$m_\alpha = \hat{m}_\alpha(\epsilon_\beta, \mu_\beta, \gamma_\beta, s, \mathbf{g}, c_a : \mathbf{X}), \tag{3.28}$$

$$\mathbf{t}_\alpha = \hat{\mathbf{t}}_\alpha(\epsilon_\beta, \mu_\beta, \gamma_\beta, s, \mathbf{g}, c_a : \mathbf{X}), \tag{3.29}$$

where α and β can vary from 1 to n. The $\mathbf{X}$ indicates a possible spatial dependence. Substituting the above equations together with eqns. (3.17) and (3.18), into eqn. (3.15) yields

$$A\dot{s} + B_i \dot{g}_i + \sum_{\alpha=1}^{n} (H1_\alpha \dot{\epsilon}_\alpha + H2_\alpha \dot{\mu}_\alpha + H3_{\alpha i} \dot{\gamma}_{\alpha i}) + \sum_{a=1}^{m} H4_a \dot{c}_a + G = 0, \tag{3.30}$$

with coefficients

$$A = -\left(\frac{\partial e}{\partial s} - \theta\right), \tag{3.31}$$

$$B_i = -\frac{\partial e}{\partial g_i}, \tag{3.32}$$

$$G = -\frac{q_i}{\theta} g_i - \xi\theta, \tag{3.33}$$

$$H1_\alpha = -\frac{\partial e}{\partial \epsilon_\alpha} + p_\alpha, \tag{3.34}$$

$$H2_\alpha = -\frac{\partial e}{\partial \mu_\alpha} + m_\alpha, \tag{3.35}$$

$$H3_{\alpha i} = -\frac{\partial e}{\partial \gamma_{\alpha i}} + t_{\alpha i}, \tag{3.36}$$

$$H4_a = -\frac{\partial e}{\partial c_a} + M_a. \tag{3.37}$$

The vectorial subscript i can vary from 1 to 3, while the doublet numbering subscript α varies from 1 to n, and the chemical subscript a varies from 1 to m. All these coefficients must be zero since

1. Neither the coefficients in eqn. (3.30) nor, by constitutive assumptions (3.23)–(3.29), the material response functions depend on $\dot{\epsilon}_\alpha$, $\dot{\mu}_\alpha$, $\dot{\gamma}_{\alpha i}$, $\dot{c}_a$, $\dot{s}$, or $\dot{g}_i$.
2. The results obtained for a particular choice of $\dot{\epsilon}_\alpha$, $\dot{\mu}_\alpha$, $\dot{\gamma}_{\alpha i}$, $\dot{c}_a$, $\dot{s}$, and $\dot{g}_i$ must hold for all choices of $\dot{\epsilon}_\alpha$, $\dot{\mu}_\alpha$, $\dot{\gamma}_{\alpha i}$, $\dot{c}_a$, $\dot{s}$, and $\dot{g}_i$.
3. Processes can be chosen in such a way to require each coefficient (3.31)–(3.37) to vanish in turn.

We find, analogous to the treatment in Chap. 2, that

$$p_\alpha = \frac{\partial e}{\partial \epsilon_\alpha}, \; m_\alpha = \frac{\partial e}{\partial \mu_\alpha}, \; t_{\alpha i} = \frac{\partial e}{\partial \gamma_{\alpha i}}, \; \theta = \frac{\partial e}{\partial s}, \text{ and } M_a = \frac{\partial e}{\partial c_a}. \tag{3.38}$$

Since e is not a function of $\mathbf{g}$ (as $B_i = 0$), q, M_a the microstresses are not functions of $\mathbf{g}$. The requirement that $G = 0$ shows that

$$\frac{q_i}{\theta} = -\xi\theta, \tag{3.39}$$

reducing the energy equation (3.15) for elastic solids to

$$-(\dot{e} - \theta\dot{s}) + \sum_{\alpha=1}^{n}(p_\alpha\dot{\epsilon}_\alpha + m_\alpha\dot{\mu}_\alpha + \mathbf{t}_\alpha \cdot \dot{\gamma}_\alpha) + \sum_{a=1}^{m} M_a\dot{c}_a = 0. \tag{3.40}$$

3.6 Linearization and Superposed Rigid Body Motions

Up to this point, our developments have been fully non-linear in nature. In the present form of DM, as presented in Chapter 1, the microstrains are defined by the geometrically linear ($|\Delta \mathbf{u}_\alpha| \ll \eta_\alpha$, $\epsilon_\alpha \ll 1$) definitions

$$\epsilon_\alpha = \frac{\boldsymbol{\tau}_\alpha^\circ \cdot \Delta \mathbf{u}_\alpha}{\eta_\alpha}, \tag{3.41}$$

$$\mu_\alpha = \frac{\boldsymbol{\tau}_\alpha^\circ \cdot \Delta \boldsymbol{\phi}_\alpha}{\eta_\alpha}, \tag{3.42}$$

$$\gamma_\alpha = -\left(\boldsymbol{\phi} + \frac{1}{2}\Delta\boldsymbol{\phi}_\alpha - \boldsymbol{\tau}_\alpha^\circ \times \boldsymbol{\tau}_\alpha\right) \times \boldsymbol{\tau}_\alpha^\circ. \tag{3.43}$$

Using these definitions, we must linearize all the balance laws. Starting with mass balance, we consider the center of mass velocity, $\mathbf{v}$, to be $\mathcal{O}(\epsilon)$ but the velocity of each component, $\mathbf{v}_a$, to be $\mathcal{O}(1)$. This approach readily leads to a restatement of eqn. (3.2)

$$\frac{\partial \rho}{\partial t} = -\text{div}(\rho \mathbf{v}), \tag{3.44}$$

from a summation of eqn. (3.1). The geometrically linearized equations of momentum are

$$\sum_{\alpha=1}^{n}\sum_{\chi=1}^{M}(-1)^{\chi+1}\frac{\eta_{\alpha}^{\chi-1}}{\chi!}\overset{\circ}{\tau}_{\alpha k_1}\cdots\overset{\circ}{\tau}_{\alpha k_\chi}\frac{\partial^{\chi}(p_{\alpha i}+t_{\alpha i})}{\partial x_{k_1}\cdots\partial x_{k_\chi}}+\sum_{a=1}^{m}\rho_a F_{ai}=\rho\frac{\partial^2 u_i}{\partial t^2}, \tag{3.45}$$

where the partial derivative has replaced the material time derivative on the right hand side. The balance equations for moment of momentum (3.11) and the reduced energy/entropy balance equation (3.40) remain unchanged in form.

Similar to the treatment presented in Chap. 2, consideration of superposed rigid body motions (SRBMs) of the form

$$u_i \rightarrow u_i^+ \equiv Q_{ij}(t)u_j + c_i(t), \tag{3.46}$$

$$\phi_i \rightarrow \phi_i^+ \equiv Q_{ij}(t)\phi_j, \tag{3.47}$$

applied to the kinematic relations of DM (3.41)–(3.43) leads to the conclusion that the DM microstrain measures are properly invariant. Similarly, significant restrictions on the functional forms of the microstresses are obtained

$$p_\alpha = \hat{p}_\alpha(\epsilon_\beta,\mu_\beta,\gamma_\beta,s,c_a:\mathbf{X}), \tag{3.48}$$

$$m_\alpha = \hat{m}_\alpha(\epsilon_\beta,\mu_\beta,\gamma_\beta,s,c_a:\mathbf{X}), \tag{3.49}$$

$$\mathbf{t}_\alpha = \sum_{\beta=1}^{n}\hat{\mathbf{I}}_{\alpha\beta}(\epsilon_\beta,\mu_\beta,\gamma_\beta,s,c_a:\mathbf{X})\boldsymbol{\gamma}_\beta, \tag{3.50}$$

$$M_a = \hat{M}_a(\epsilon_\beta,\mu_\beta,\gamma_\beta,s,c_a:\mathbf{X}), \tag{3.51}$$

where γ_β is the magnitude of $\boldsymbol{\gamma}_\beta$.

Restrictions for the other functions in our development are also obtained

$$e = \hat{e}(\epsilon_\alpha,\mu_\alpha,\gamma_\alpha,s,c_a:\mathbf{X}), \tag{3.52}$$

$$\theta = \hat{\theta}(\epsilon_\alpha,\mu_\alpha,\gamma_\alpha,s,c_a:\mathbf{X}), \tag{3.53}$$

$$\xi = \hat{\xi}(\epsilon_\alpha,\mu_\alpha,\gamma_\alpha,s,c_a:\mathbf{X}), \tag{3.54}$$

$$\mathbf{q} = \sum_{\alpha=1}^{n}\hat{\mathbf{H}}_\alpha(\epsilon_\beta,\mu_\beta,\gamma_\beta,s,c_a,g,\boldsymbol{\gamma}_\beta\cdot\mathbf{g}:\mathbf{X})\boldsymbol{\gamma}_\alpha+\cdots + \hat{\mathbf{K}}(\epsilon_\beta,\mu_\beta,\gamma_\beta,s,c_a,g,\boldsymbol{\gamma}_\beta\cdot\mathbf{g}:\mathbf{X})\mathbf{g}, \tag{3.55}$$

where g is the magnitude of grad(θ). Again, as in Chap. 2, the dependence of $\mathbf{q}$ on $\boldsymbol{\gamma}_\alpha$ can be removed in the linear theory.

3.7 Linear Elastic Doublet Mechanics

Assuming a physically linear constitutive relation between the microstrains and microstresses, we obtain

$$p_\alpha = \sum_{\beta=1}^{n}(A_{\alpha\beta}\,\epsilon_\beta + B_{\alpha\beta}\,\mu_\beta + C_{\alpha\beta}\,\gamma_\beta) + J_\alpha \Lambda + \sum_{a=1}^{m} M1_{\alpha a}\,c_a, \quad (3.56)$$

$$m_\alpha = \sum_{\beta=1}^{n}(D_{\alpha\beta}\,\epsilon_\beta + E_{\alpha\beta}\,\mu_\beta + F_{\alpha\beta}\,\gamma_\beta) + K_\alpha \Lambda + \sum_{a=1}^{m} M2_{\alpha a}\,c_a, \quad (3.57)$$

$$t_{\alpha i} = \sum_{\beta=1}^{n} I_{\alpha\beta ij}\,\gamma_{\beta j}, \quad (3.58)$$

$$M_a = \sum_{\alpha=1}^{n}(G_{\alpha a}\epsilon_\alpha + H_{\alpha a}\mu_\alpha + R_{\alpha a}\gamma_\alpha) + L_a \Lambda + \sum_{b=1}^{m} M3_{ab}\,c_b, \quad (3.59)$$

$\Lambda = s - s^\circ$ is an increment of the entropy per unit volume and s° is the entropy per unit volume of the solid in an initial state.

The continuity of e implies the independence of the order of partial differentiation, i.e.,

$$\frac{\partial^2 e}{\partial\epsilon_\alpha \partial\epsilon_\beta} = \frac{\partial^2 e}{\partial\epsilon_\beta \partial\epsilon_\alpha} = \frac{\partial p_\alpha}{\partial\epsilon_\beta} = \frac{\partial p_\beta}{\partial\epsilon_\alpha} = A_{\alpha\beta} = A_{\beta\alpha}, \quad (3.60)$$

leads to the conclusions that

$$\left.\begin{array}{lll} A_{\alpha\beta} = A_{\beta\alpha}, & E_{\alpha\beta} = E_{\beta\alpha}, & I_{\alpha\beta ij} = I_{\beta\alpha ji}, \\ B_{\alpha\beta} = D_{\beta\alpha}, & G_{\alpha a} = M1_{\alpha a}, & H_{\alpha a} = M2_{\alpha a}, \\ M3_{ab} = M3_{ba}, & C_{\alpha\beta} = F_{\alpha\beta} = R_{\alpha a} = 0. & \end{array}\right\} \quad (3.61)$$

As can be seen from this exercise in mathematical symmetry, it is wise to group the chemical potential, M_a, with the microstresses, as we have done. Using relations (3.61), relations (3.48)–(3.51) are rewritten as

$$p_\alpha = \sum_{\beta=1}^{n}(A_{\alpha\beta}\,\epsilon_\beta + B_{\alpha\beta}\,\mu_\beta) + J_\alpha \Lambda + \sum_{a=1}^{m} G_{\alpha a}\,c_a, \quad (3.62)$$

$$m_\alpha = \sum_{\beta=1}^{n}(D_{\alpha\beta}\,\epsilon_\beta + E_{\alpha\beta}\,\mu_\beta) + K_\alpha \Lambda + \sum_{a=1}^{m} H_{\alpha a}\,c_a, \quad (3.63)$$

$$t_{\alpha i} = \sum_{\beta=1}^{n} I_{\alpha\beta ij}\,\gamma_{\beta j}, \quad (3.64)$$

$$M_a = \sum_{\alpha=1}^{n}(G_{\alpha a}\epsilon_\alpha + H_{\alpha a}\mu_\alpha) + L_a \Lambda + \sum_{b=1}^{m} M3_{ab}\,c_b, \quad (3.65)$$

with the conditions $A_{\alpha\beta} = A_{\beta\alpha}$, $E_{\alpha\beta} = E_{\beta\alpha}$, $I_{\alpha\beta ij} = I_{\beta\alpha ji}$, $B_{\alpha\beta} = D_{\beta\alpha}$, and $M3_{ab} = M3_{ba}$.

3.8 Closure

From the preceding analysis it is concluded that:

1. The internal energy e is *not* a function of the temperature gradient $\mathbf{g} = \mathrm{grad}(\theta)$.
2. The microstresses and microstrains obey the following relations

$$p_\alpha = \frac{\partial e}{\partial \epsilon_\alpha}, \quad m_\alpha = \frac{\partial e}{\partial \mu_\alpha}, \quad t_{\alpha i} = \frac{\partial e}{\partial \gamma_{\alpha i}}, \quad M_a = \frac{\partial e}{\partial c_a}, \tag{3.66}$$

   and thus the microstresses are also independent of $\mathrm{grad}(\theta)$.
3. The internal energy e and the temperature θ are related by

$$\theta = \frac{\partial e}{\partial s}. \tag{3.67}$$

4. The microstrains: ϵ_α, μ_α and $\boldsymbol{\gamma}_\alpha$ are properly invariant under arbitrary finite rotations:

$$\epsilon_\alpha^+ = \epsilon_\alpha, \qquad \mu_\alpha^+ = \mu_\alpha, \qquad \boldsymbol{\gamma}_\alpha^+ = \mathbf{Q}\boldsymbol{\gamma}_\alpha. \tag{3.68}$$

5. For an elastic solid, the functional representations of the microstresses must obey the following relations:

$$p_\alpha = \hat{p}_\alpha(\epsilon_\beta, \mu_\beta, \gamma_\beta, s, c_a : \mathbf{X}), \tag{3.69}$$

$$m_\alpha = \hat{m}_\alpha(\epsilon_\beta, \mu_\beta, \gamma_\beta, s, c_a : \mathbf{X}), \tag{3.70}$$

$$\mathbf{t}_\alpha = \sum_{\beta=1}^{n} \hat{\mathbf{I}}_{\alpha\beta}(\epsilon_\beta, \mu_\beta, \gamma_\beta, s, c_a : \mathbf{X})\boldsymbol{\gamma}_\beta, \tag{3.71}$$

$$M_a = \hat{M}_a(\epsilon_\beta, \mu_\beta, \gamma_\beta, s, c_a : \mathbf{X}), \tag{3.72}$$

   where γ_β is the magnitude of $\boldsymbol{\gamma}_\beta$.
6. The constitutive laws for a homogeneous linear elastic solid reduce to

$$p_\alpha = \sum_{\beta=1}^{n}(A_{\alpha\beta}\,\epsilon_\beta + B_{\alpha\beta}\,\mu_\beta) + J_\alpha \Lambda + \sum_{a=1}^{m} G_{\alpha a}\,c_a, \tag{3.73}$$

$$m_\alpha = \sum_{\beta=1}^{n}(D_{\alpha\beta}\,\epsilon_\beta + E_{\alpha\beta}\,\mu_\beta) + K_\alpha \Lambda + \sum_{a=1}^{m} H_{\alpha a}\,c_a, \tag{3.74}$$

$$t_{\alpha i} = \sum_{\beta=1}^{n} I_{\alpha\beta ij}\,\gamma_{\beta j}, \tag{3.75}$$

$$M_a = \sum_{\alpha=1}^{n}(G_{\alpha a}\epsilon_\alpha + H_{\alpha a}\mu_\alpha) + L_a \Lambda + \sum_{b=1}^{m} M3_{ab}\,c_b, \tag{3.76}$$

   with the conditions $A_{\alpha\beta} = A_{\beta\alpha}$, $E_{\alpha\beta} = E_{\beta\alpha}$, $I_{\alpha\beta ij} = I_{\beta\alpha ji}$, $B_{\alpha\beta} = D_{\beta\alpha}$, and $M3_{ab} = M3_{ba}$.

4. Comparison with Other Theories

V. T. Granik

4.1 Introduction

As is clear from the previous chapters, doublet mechanics (DM) is a new theory of elasticity stemming from and based on the consideration of the discrete microstructure of solids. Doublet mechanics is not the only microstructural model—there are many similar theories that have been developed during the last 30–35 years. The existance of a wide variety of microstructural theories of elasticity leads to the inevitable question: what are the salient features of DM that distinguish it from the others? In this chapter we discuss this question with regard to well-known mathematical models of elasticity (WME).

Since some elastic solids with microstructure (crystals) are studied in lattice dynamics (LD), it is instructive to include LD in our analysis as well. Given the goal of this chapter, our consideration of LD is concise. We only touch upon those peculiarities of LD that make it radically different from DM. We begin with a brief discussion of LD and then go over the WME. By comparing LD and the WME on the one hand, and DM on the other hand, we determine some domains of applicability of DM.

4.2 Lattice Dynamics

4.2.1 General Remarks

The fundamentals and basic applications of LD are considered in the monographs by Born and Huang (1962), Maradudin et al (1971), Böttger (1983) and Askar (1985). Elements of LD are also treated in books on solid state physics by Landsberg (1969), Cochran (1973), Elliott and Gibson (1974), Venkatamaran (1975), Rosenberg (1975), Ashcroft and Mermin (1976) and Kittel (1986) to name but a few. Various applications of LD are dealt with in a vast number of papers.

Lattice dynamics is concerned with the motion of the nuclei (and more closely bound electrons) in crystals and may be divided into two branches. The first one is based on the adiabatic approximation (Born and Oppenheimer (1927) that allows one to separate the motion of nuclei from that of electrons. The motion of electrons is studied within the framework of quantum mechanics. The motion of separated ions in a crystal lattice is considered by the adiabatic version of LD in which only the ion-ion interaction is taken into account and the electron motion is ignored. It is worth mentioning that, despite its wide applicibility, this version of LD is not completely infallible and gives rise to critical discussions (Ziman 1960, Haug 1972).

The second branch of LD does not apply the Born-Oppenheimer approximation and is therefore able to study more subtle and complex phenomena caused not only by the ion-ion interactions but also by the ion-electron interactions (see, e.g., Böttger (1983, pp. 59–62)).

Doublet mechanics is, in a sense, closer to the adiabatic version of LD. Therefore in what follows we mean the adiabatic branch of LD and call it simply LD. Using the above quoted sources, we can outline some salient features of LD that are important in comparison with those of DM.

4.2.2 Salient Features

Lattice dynamics studies vibrations of the atomic nuclei of solid crystals, the nuclei being considered as material points ("particles") mutually bonded by elastic interatomic forces. The particle vibrations can be small or large compared with interatomic distances. The corresponding approximations of the theory are called harmonic and anharmonic, respectively. The harmonic approximation leads to linear governing equations and is usually referred to as the classical LD (Madelung 1978). We treat below the classical LD only.

Dealing with crystals, LD makes crucial use of their translational symmetry. Due to the symmetry involved, the governing equations of LD are simplified almost to the extreme. Such a great benefit is achieved with certain losses. Let us quote to the point (Landsberg 1969, pp. 331–332):

> The advantages of translational symmetry can only be fully obtained in an infinite crystal. On the other hand in order that, for instance, the energy shell be finite, a finite crystal is desirable. Also it would be convenient if in some way the group of translational symmetry operations could be made finite, since the theory of finite groups is simpler than that of infinite groups. All these ends can be achieved by the device of periodic boundary conditions.

LD is based on the following far-reaching, but to a certain extent artificial, premises:

1. Any crystal is an *infinite* lattice structure.
2. The crystal obeys some *devised* periodic boundary conditions (PBC).

How do these concepts work in LD? Since a crystal is assumed to be infinite LD is only capable of studying a particular case of the atom vibrations, namely, the propagation of body waves. The waves being determined by the angular frequencies ω and the wave vectors $\mathbf{k}$. Lattice dynamics is primarily aimed at solving various eigenvalue problems, i.e., finding functions $\omega(\mathbf{k})$ by considering the above body waves on the basis of the linear governing equations. There are also other problems treated by LD such as the frequency spectrum $g(\omega) = 2\omega G(\omega^2)$ which is most commonly applied to the calculation of thermodynamic properties of crystals. However, this problem is not quite independent because there are certain relations between $G(\omega^2)$ and $\omega(\mathbf{k})$ based on the Bowers-Rosenstock equation (Maradudin et al 1971).

The infinite crystal has an infinite number of atoms and hence must have an infinite number of eigenvalues. But the crystal's infinite expanse, along with its symmetry, bring about a crucial simplification of the problem: they reduce the number of eigenvalues to a finite number by considering only a finite number of ions in a unit cell of the underlying lattice.

Despite, however, the finite number of eigenvalues, there remains some uncertainty in the values the vector $\mathbf{k}$ can assume. This obstacle is overcome by the aforementioned periodic boundary condition (PBC) first proposed by Born and von Kármán (1912). The PBC postulates that ionic displacements are periodic with the periodicity of a certain characteristic length L. Due to the PBC, the allowed values of $\mathbf{k}$ lie in one cell of the reciprocal lattice of the crystal, i.e., in the first Brillouin zone. The PBC also simplifies the study of the frequency spectra $g(\omega)$.

A remark should be made about the characteristic length L. This parameter is not the length nor the width of a real finite crystal. It is a *conditional* quantity to be determined with regard to the problem in question. We thus see that the PBC has nothing in common with the *real* boundary conditions well-known from mathematical physics, continuum mechanics, etc. The PBC applications are rigorously justified under two limitations:

1. The outside surface of a crystal is free from forces and
2. Interionic forces are short-range; they may also be long-range but not including all the ions of a crystal.

In general, the role of the PBC in LD is not restricted to placing all the values of $\mathbf{k}$ in the first Brillouin zone and simplifying the calculations of frequency spectra. There is another—crucial—reason for incorporating such a mathematical device as the PBC in LD (Maradudin et al 1971, p. 38):

> If the calculations to be described in this book (on LD) were critically dependent on the particular choice of boundary conditions it would be virtually impossible to develop a theory with any degree of generality. Fortunately this does not turn out to be the case, and the very simplest boundary conditions first proposed by Born and von Karman may be used to develop a theory with as much validity as the physical axioms permit.

4.2.3 Domains of Inapplicability

We see that LD can exist as a *general* theory only on the basis of a particular boundary condition (the PBC) and it is, in general, incompatible with arbitrary boundary conditions. It follows that LD is incapable of solving those vibration problems where boundary conditions differ from the PBC with its strong requirement that no forces be applied to the outside surface of a crystal. Meanwhile, in numerous cases that are intrinsic to the wide-ranging problems of elasticity, quite an opposite situation prevails when certain forces act on the bounding surface of a solid body, which is not necessarily a crystal, or the surface is subjected to specified displacements. Thus, LD is, in general, inapplicable to the boundary problems of elasticity. There are also other limitations. Since LD was developed to study only elastic vibration it is unable to address both inelastic and static problems of solid mechanics.

Despite being an efficient theory for crystal vibrations, LD is not a universal tool to solve all the problems of deformable solids. There are extensive domains where LD is inapplicable. Let us mention some of these.

1. Quasielastic and inelastic behavior of solids
   - Viscoelasticity[1]
   - Plasticity
2. Dynamic boundary problems
   - Vibrations of soil bases (foundations)
   - Propagation of surface waves[2,3]
   - Stress concentration
     - ☐ Around holes, cavities, notches, and the like
     - ☐ Around inclusions
     - ☐ Under punches, beams, shells, etc.
     - ☐ At the tips of cracks
3. Static boundary problems
   - Strains and stresses in soil bases (foundations)[4]
   - Stress concentration
     - ☐ Around holes, cavities, notches, and the like
     - ☐ Around inclusions
     - ☐ Under punches, beams, shells, etc.
     - ☐ At the tips of cracks
4. Failure criteria[1]
5. Fracture mechanics

[1] These problems are touched upon within the present book on the basis of DM.
[2] These problems are dealt with in the present book on the basis of DM.
[3] In solids with independent rotations ϕ of underlying particles. In a particular case of central interatomic forces when $\phi \equiv 0$ some problems of surface waves can be treated by means of LD. The treatment is excessively complicated since in general such problems are not inherent in LD (see, e.g., Maradudin et al (1971, pp. 520–582)). For comparison with DM, see Chap. 7 in this book).
[4] Including the Flamant problem which is considered here in Chap. 8.

4.3 Doublet Mechanics

4.3.1 Salient Features

The fundamentals of DM are stated in detail in the previous chapters. Here we are going to recount and reemphasize only those salient features of DM that are important in comparison with other theoretical models of elasticity.

From Chap. 1 it follows that doublet mechanics is not a deductive phenomenological theory stemming only from purely axiomatic principles. DM is an inductive mathematical model based on a certain physical prototype. This prototype is, in turn, a discrete physical model of some real solids with a pronounced microstructure. DM has been developed according to the stages shown in Fig. 4.1.

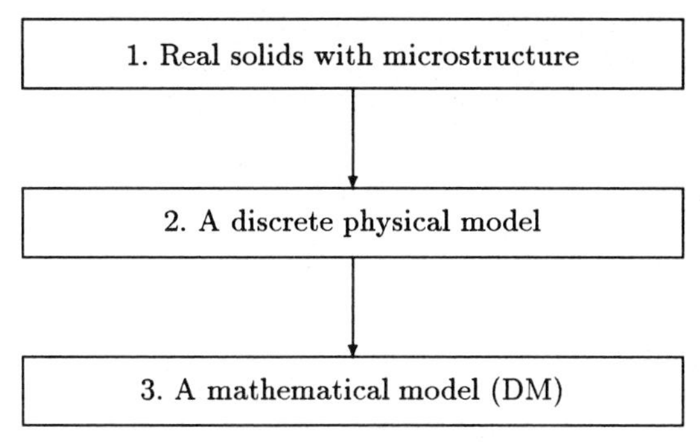

Fig. 4.1. Stages of developing DM

The first element in Fig. 4.1—the real solid in question—may be a crystal or a granular array or the Earth's crust, etc., whose underlying material units—atoms, grains, tectonic plates—are considerably smaller than the corresponding solid as a whole. Due to this feature the material units may be called particles. The arrangement of the particles in a solid body, i.e., its microstructure, may vary from regular and almost regular (quasi-regular) to completely irregular (chaotic). For example, a perfect crystal has a regular microstructure, an array of tectonic plates of the Earth's crust spreading over tens of kilometers usually possesses an irregular microstructure, whereas granular packings may have any microstructure, from regular and quasi-regular to chaotic, depending on their genesis.

According to Chap. 1, the second element in Fig. 4.1—a discrete physical model of DM—is a set of numerous identical small elastic spheres whose centers form a regular microstructure, a Bravais lattice with the coordination number m and the valence $n = m/2$. The couples of adjacent particles A and B_α called doublets create elementary blocks of the discrete microstructure, their length being η_α, $\alpha = 1, 2, \ldots, n$. The kinematics of the doublets is defined by mutually independent particle translations $\mathbf{u}$ and rotations $\boldsymbol{\phi}$. When passing in a doublet from one particle A to another B_α, the vectors $\mathbf{u}$ and $\boldsymbol{\phi}$ change by $\Delta\mathbf{u}_\alpha$ and $\Delta\boldsymbol{\phi}_\alpha$, respectively. The vectors $\boldsymbol{\phi}$, $\Delta\mathbf{u}_\alpha$ and $\Delta\boldsymbol{\phi}_\alpha$ engender the doublet microstrains of elongation, shear, and torsion, as well as the corresponding doublet microstresses of the same nomenclature. In general, any particle may interact with any of the others. However, in the present version of DM the interparticle microforces and microcouples are assumed to be in a sense local[5], i.e., confined as a rule to nearest and sometimes to the next-nearest and third-nearest neighbors. Unlike LD, the microforces in DM are not necessarily central because of possible microshears.

The third element in Fig. 4.1—a mathematical model—is based on the assumption that the spatial increments $\Delta\mathbf{u}_\alpha$ and $\Delta\boldsymbol{\phi}_\alpha$ may be expanded into infinite three-dimensional Tailor series. The main consideration is focused on arbitrary truncations of these series that include the first $M \geq 1$ spatial gradients $\partial^k\mathbf{u}$ and $\partial^k\boldsymbol{\phi}$ $(k = 1, 2, \ldots, M)$. The truncated series result in the kinematic equations of DM in which the doublet microstrains are expressed via the gradients $\partial^k\mathbf{u}$ and $\partial^k\boldsymbol{\phi}$ up to the Mth order. In view of the kinematic equations, the variational principle of virtual work brings about a coherent set of the motion/equilibrium equations and the natural boundary conditions written in terms of the doublet microstresses.

The doublet microstrains and microstresses constitute the "language" of DM. It should be emphasized that the doublet microstresses have a clear physical meaning They represent the elastic microforces and microcouples of interaction between any two particles of a doublet. As a consequence, on the basis of the doublet microstresses, one is capable of revealing such phenomena that are inaccessible by means of the conventional macrostresses (see, for example, the Flamant problem in Chap. 8). Being either scalars or vectors, the doublet microstresses as mathematical quantities are much simplier than the conventional macrostresses (second-rank tensors). This quality considerably facilitates the formulation of invariant constitutive equations. As a whole, the novel concept of doublet microstrains and microstresses plays a crucial role in making DM different from, and competitive with, the other theories of elasticity (see, below, Sect. 4.4 and Sect. 4.5).

Let us return to the governing equations of DM. These relations have the following distinctive features:

1. They are differential.
2. They include the scale parameters η_α.
3. The parameters η_α enter into the equations in an explicit form, as the multipliers (η_α^{k-1})[6] $(k = 1, 2, \ldots, M)$ of corresponding equation terms.

[5] This term has also another meaning specified below (see Sect. 4.4.1) when applied to *local* theories of elasticity as the opposites to *nonlocal* ones.

[6] η_α^k for the microshears.

4. The higher the level of approximation M, the closer DM is to its discrete physical prototype. A case of large values of M is exemplified below, in Chap. 5.
5. If we let $M \to \infty$, the differential governing equations can be transformed into the consequent finite difference equations in spatial variables. With these equations, DM becomes an exact theory of elasticity of the underlying discrete physical model. In Chap. 7 we will consider an example of the derivation and the application of such difference governing equations.

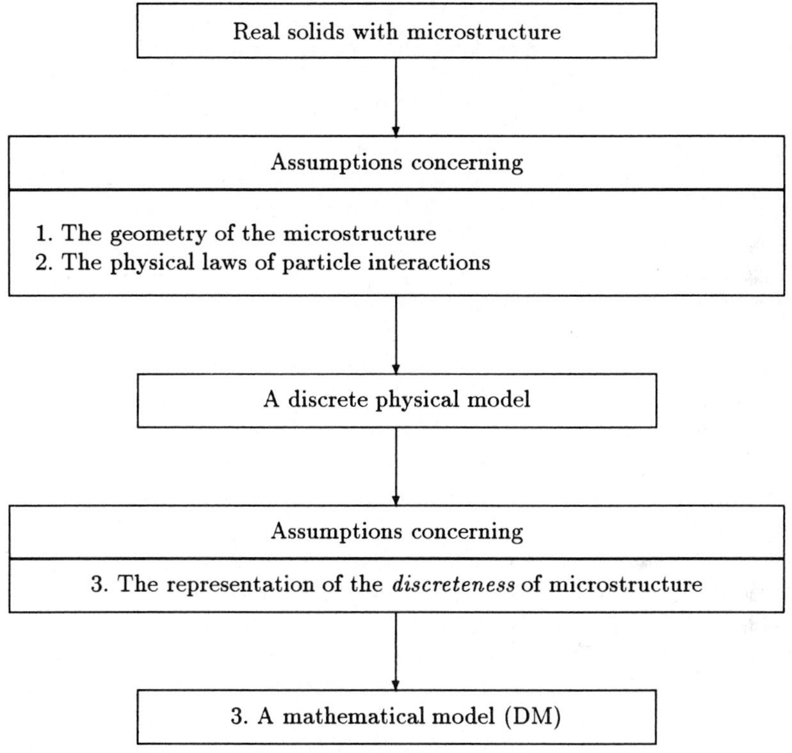

Fig. 4.2. An extended pattern of developing DM

Figure 4.1 and the above remarks show that the road from a real solid with microstructure to its mathematical model, DM, passes through the set of three basic assumptions pointed out in Fig. 4.2. As in all the mathematical models of elasticity, there are certain differences between the fundamentals of microstructure intrinsic to real solids (they are italicized in Fig. 4.2) and those assumed in DM. From Fig. 4.2 it follows that DM enables one to get an exact mathematical description of the mechanical behavior of a solid with microstructure if the two necessary conditions are satisfied:

1. A solid in question and its discrete physical prototype coincide.
2. Doublet mechanics is applied in the version with the difference governing equations of kinematics and statics/dynamics that hold both for the interior and the boundaries of a solid.

Departures from these conditions lead to some discrepancies between theoretical quantities—microstrains, microstresses, elastic translations and microrotations—and their true values. Such discrepancies are different in nature from the errors of measurements, calculations, etc., and might therefore be called the errors of a mathematical model (EMM). According to the above necessary conditions, the set of the EMM in doublet mechanics is made up of two subsets: $\{\text{EMM}\}=\{\text{EMM-1}\}\cup\{\text{EMM-2}\}$, $\{\text{EMM-1}\}\cap\{\text{EMM-2}\}=\emptyset$. The subset $\{\text{EMM-1}\}=\emptyset$ if the first condition is obeyed, and $\{\text{EMM-2}\}=\emptyset$ if the second condition is satisfied. In the latter case, as noted previously, doublet mechanics is an exact theory of elasticity of the underlying discrete physical model. The notion of the EMM is useful in the comparison of DM and other theories of elasticity.

4.3.2 Comparison with Classical and Cosserat's Theories of Elasticity

Although all the governing equations and boundary conditions of DM are derived and written in terms of the doublet microstrains and microstresses, generalized macrostresses—"force" $\sigma_{ij}^{(M)}$ and "couple" $M_{ij}^{(M)}$—are also introduced. In a first approximation of the theory ($M = 1$), they coincide with the conventional Cauchy σ_{ij} and Cosserat M_{ij} stress tensors; at $M \geq 2$ they have no counterparts.

The generalized macrostresses $\sigma_{ij}^{(M)}$ and $M_{ij}^{(M)}$ can be computed via the doublet microstresses by two formulas (1.40) and (1.41) which are briefly referred to as the macro-microrelations. The macro-microrelations have an essential feature, if $M = 1$ their inversion is, in general, impossible since the number of doublet microstresses is usually more than that of the macrostresses: σ_{ij} and M_{ij}. However, if we know the doublet microstresses we *always* are able to calculate the unknown macrotensors σ_{ij} and M_{ij}. If we know σ_{ij} and M_{ij}, we are, in general, unable to determine the doublet microstresses. This means that the doublet microstresses are more informative than the Cauchy σ_{ij} and Cosserat M_{ij} macrostresses and they provide, therefore, deeper insight into the mechanical behavior of solids.

The above macro-microrelations enable one to express the equations of motion of DM in terms of both the doublet microstresses and the generalized macrostresses $\sigma_{ij}^{(M)}$ and $M_{ij}^{(M)}$. At $M = 1$, the latter relations reduce to the differential equations of motion of either the classical or the Cosserat elastic continua depending on whether the couple macrostresses M_{ij} are respectively equal to or different from zero.

As follows from eqn. (1.41), Cosserat's couple macrostresses M_{ij} vanish in two cases:

1. The doublet microstresses of torsion $\mathbf{m}_\alpha$ and shear $\mathbf{t}_\alpha$ are equal to zero. As a result, the doublet governing equations of motion and the boundary conditions do not contain the scale parameters η_α.
2. The doublet microstresses $\mathbf{m}_\alpha \equiv 0$, but $\mathbf{t}_\alpha \not\equiv 0$. The scale parameters $\eta_\alpha \equiv 0$ and are therefore eliminated from the doublet governing equations of motion and from the boundary conditions.

These cases become possible if, respectively:

1. The particle interactions are central (axial). The solid micro- structure may therefore be called axial. It is also nonpolar ($\mathbf{m}_\alpha \equiv 0$) and explicitly scaleless ($\eta_\alpha \equiv 0$).
2. The underlying particles interact not only by central, but also by shear microforces. Hence the solid microstructure is nonaxial. As in the previous case, it is also nonpolar and explicitly scaleless.

Just for such solids—*nonpolar* and *explicitly scaleless*—the first approximation of DM reduces to classical elasticity; otherwise this approximation of DM reduces to the Cosserat continuum. We thus see that the classical theory of elasticity is explicitly (and implicitly, as shown below) scaleless whereas Cosserat's theory is scale dependent, either explicitly or implicitly. The original Cosserat model (Cosserat and Cosserat 1907, Cosserat and Cosserat 1909) is implicitly scaling. As to the explicit scale version of Cosserat's continuum its existence was first established by Granik and Ferrari (1993).

4.4 Micromechanics

4.4.1 Classification

All the theories of elastic solids with microstructure are referred to as micromechanics. The distinctive feature of micromechanics is nonlocality. This means that any micromechanical theory incorporates—explicitly or implicitly—some scale parameters inherent in the solid under consideration. The parameters may be the sizes and the separations of the underlying particles, the dimensions of internal microstructural cells, characteristic ranges of particle interactions, and so on. However, whatever the scale parameters, they are assumed to be small when compared with the total dimensions of the solid. By this definition and according to Sect. 4.3.1 and Sect. 4.3.2, doublet mechanics and Cosserat's model are nonlocal theories and belong to micromechanics, whereas classical elasticity does not.

As in the case of lattice dynamics, we will further consider only those nonlocal theories of elasticity that are based on the harmonic approximation (see Sect. 4.2.2) and include therefore only linear governing equations as DM does. For this reason we avoid the so-called Toda systems, i.e., microstructural models based on regular lattices with exponential nearest neighbor interactions (Toda 1988). Therefore, talking about micromechanics, we mean only linear microstructural theories.

Table 4.1. Classification of Micromechanical Theories

No.	Theories	Governing Equations	Nonlocality	Stresses
1.	DM	Differential	Explicit	Doublet microstresses
2.	DM	Difference	Explicit	Doublet microstresses
3.	Other theories	Differential	Implicit	Tensor macrostresses
4.	Other theories	Integral	Explicit	Tensor macrostresses

Despite a wide variety of the micromechanical theories, they might be classified with regard to the principal features of (i) governing equations, (ii) underlying nonlocality, and (iii) underlying stresses, as shown in Table 4.1. The differences in these three attributes make DM distinct from and competitive with the other microstructural models. To clear up the point we should take a look at the other nonlocal theories keeping in mind the classification given in Table 4.1.

4.4.2 Nonlocal Theories of Differential Type

General Remarks. This type of nonlocal theories (No. 3 in Table 4.1) includes many models developed during a relatively short time, from the end of the 1950s to the early 1970s, mainly by Ericksen and Truesdell (1958), Günther (1958), Aero and Kuvshinski (1960), Grioli (1960), Toupin (1962, 1964, 1968), Mindlin and Tiersten (1962), Dahler and Scriven (1963), Koiter (1964), Mindlin (1964, 1965), Green and Rivlin (1964a), Green (1965), Eringen (1966, 1968), Jaunzemis (1967, pp. 223–250, 394–425), Rivlin (1968) and Stojanović (1972). Being different in many details, these theories have, nonetheless, one common principal characteristic. They all take into account not only the first order spatial gradients of displacements (as in classical elasticity) and those of microrotations (as in Cosserat's model), but also the higher order gradients of the kinematic variables. From this viewpoint, such models generalize the Cosserat medium, as well as the classical one, and are therefore referred to as generalized Cosserat continua (Teodorescu and Soós 1973). In special instances these continua are also called gradient, structured, oriented, multipolar, micromorphic, Cosserat-type, materials of grade N, etc.

Due to the higher order gradients of kinematic variables, scale parameters are involved in governing equations of the generalized Cosserat theories (GCT) and thus impart to them nonlocal character. But in contrast with DM, all geometrical parameters of microstructure—scale and non-scale ones—enter into the GCT *implicitly* (see Table 4.1). This feature of the GCT results in far-reaching consequences and is therefore considered below. In order to highlight the point and avoid insignificant details, we adopt the following simplifications which do not affect principal differences between the GCT and DM:

1. The analysis is carried out in a fixed rectangular Cartesian frame of reference $\{x_i\}$, $i = 1, 2, 3$, so that there is no difference between covariant and contravariant spatial derivatives.
2. A solid in question is assumed to be homogeneous.
3. Thermal phenomena and entropy are neglected.
4. Independent microrotations are not included and the elastic motion of a solid is described by a displacement field $\mathbf{u}$, as in classical elasticity and DM. Such a technique is, in general, tantamount to another one used in most GCT where, instead of the field $\mathbf{u}$, a field of current position vectors $\mathbf{x}$ is employed.

Scaling Role of Displacement Gradients. In the classical anisotropic elasticity, the energy of deformation E depends on the first order spatial gradients $u_{i,j} \equiv \partial u_i / \partial x_j$ of the displacements $u_i(x_k, t)$:

$$\mathcal{E} = \int_V W(u_{i,j})\, dV \equiv \frac{1}{2} \int E_{ijkl}\, u_{i,j}\, u_{k,l}\, dV \tag{4.1}$$

where t is time, V is the volume of a solid, $W(u_{i,j}) \equiv \frac{1}{2} E_{ijkl} u_{i,j} u_{k,l}$ is the energy density, E_{ijkl} is the fourth-rank tensor of elastic constants. By analogy with (Toupin 1962), we assume the existence of a material parameter μ_1 having the physical dimension of stress, $[\mathrm{ML}^{-1}\mathrm{T}^2]$, and a second material parameter μ_2 having the physical dimension of length, [L]. Equation (4.1) then takes the form

$$\mathcal{E} = \frac{1}{2} \int_V \mu_1\, \hat{E}_{ijkl}\, u_{i,j}\, u_{k,l}\, dV \tag{4.2}$$

where $\hat{E}_{ijkl}$ are dimensionless elastic constants. In view of eqns. (4.1) and (4.2), Hamilton's principle for independent variations δu_i brings about the well-known differential equations of motion of classical anisotropic elasticity (Landau and Lifshitz 1959)

$$\rho\, \ddot{u}_i = E_{ijkl}\, u_{k,jl} \equiv \mu_1\, \hat{E}_{ijkl}\, u_{k,jl}, \tag{4.3}$$

in which ρ is the mass density and $\ddot{u}_i \equiv \partial^2 u_i / \partial t^2$. The equations of motion (4.3) do not contain—explicitly or implicitly—any parameters with dimension length [L]. This means that classical elasticity is a scaleless (local) theory.

The locality of classical elasticity stems from the underlying expression (4.1) according to which the energy density W depends only on the first order displacement gradients $u_{i,j}$. This fact suggests that a theory can be made nonlocal by taking into account not only the first but also the second and the higher order spatial gradients $u_{i,j}$, $u_{i,jk}, \ldots$. Just such a postulate is adopted in the GCT. Due to the postulate, eqn. (4.1) is represented in the generalized form

$$\mathcal{E} = \int_V W(u_{i,k_1}, u_{i,k_1 k_2}, \ldots, u_{i,k_1 k_2 \ldots k_N})\, dV \tag{4.4}$$

Following Noll's terminology (Noll 1958), the materials described by the underlying eqn. (4.1) are called simple materials. On the other hand, the materials pertaining to the more complicated eqn. (4.4) are not simple. If N is the order of the highest gradient presented as an argument of the energy density, the material is called a material of grade N. It follows that a simple material is a material of grade 1, whereas nonsimple ones are of grade 2 and higher. In other words, *a scaleless theory* we may write symbolically the following chain of equivalencies: *scaleless theories* = *local theories* = *theories of continua without microstructure* = *theories of simple materials* = theories of materials of grade 1. On the contrary, *theories of materials of grades 2 and more (the GCT)* = *scale (nonlocal theories.* Let us prove these equivalences.

General features of the GCT are illustrated sufficiently well by materials of grade $N = 2$. If $N > 2$ then the analysis becomes much more complicated; these details are unessential here. Therefore, for the sake of simplicity, we restrict ourselves to materials of grade $N = 2$. Since we consider only linear elastic solids the generalized eqn. (4.4) takes a quadratic form analogous to eqn. (4.1):

$$\mathcal{E} = \frac{1}{2} \int_V \left[E_{ijkl}\, u_{i,j}\, u_{k,l} + 2E_{ijklp}\, u_{i,j}\, u_{k,lp} + E_{ijklpq}\, u_{i,jk}\, u_{l,pq}\right] dV \tag{4.5}$$

where E_{ijklp}, E_{ijklpq} are additional macroscopic tensors of elastic constants. By employing the above material parameters μ_1 and μ_2, we can represent eqn. (4.5) as follows:

$$\mathcal{E} = \frac{1}{2} \int_V \mu_1 \Big[\hat{E}_{ijkl}\, u_{i,j}\, u_{k,l} + 2\mu_2 \hat{E}_{ijklp}\, u_{i,j}\, u_{k,lp} + (\mu_2)^2 \hat{E}_{ijklpq}\, u_{i,jk}\, u_{l,pq}\Big]\, dV \tag{4.6}$$

Here $\hat{E}_{ijklp}$ and $\hat{E}_{ijklpq}$ are additional dimensionless elastic constants. On the basis of eqn. (4.6), Hamilton's principle for independent variations δu_i yields the following equations of motion typical of the GCT:

$$\rho \ddot{u}_i = E_{ijkl} u_{k,jl} + (E_{ijklp} - E_{kpijl})\, u_{k,jlp} - E_{ijklpq} u_{i,jkpq} \tag{4.7}$$

$$\equiv \mu_1 \Big[\hat{E}_{ijkl} u_{k,jl} + \mu_2 \left(\hat{E}_{ijklp} - \hat{E}_{kpijl}\right) u_{k,jlp} - (\mu_2)^2 \hat{E}_{ijklpq} u_{i,jkpq}\Big] \tag{4.8}$$

Equations (4.7) and (4.8) are drastically different from eqns. (4.3) in that they incorporate not only a parameter μ_1 of dimension of stress, but also a parameter μ_2 of dimension length. This feature makes the GCT scale (nonlocal) theories. We thus see that through the energetically conjugated elastic macrotensors E_{ijklp} and E_{ijklpq} of rank five and more the second-order displacement gradients $u_{i,jk}$ introduce nonlocality into the GCT. The same is true for the higher order gradients of u_i.

Comparison with Doublet Mechanics. Under some simplifying assumptions of no consequence here, the equations of motion of DM take the form (see Chap. 5)

$$\rho \ddot{u}_i = C_{ijkl}u_{k,jl} + C_{ijklpq}u_{l,jkpq} + C_{ijklpqrs}u_{p,jklqrs} + \cdots \tag{4.9}$$

where C_{ijkl}, C_{ijklpq}, ... are macroscopic elastic tensors of an underlying discrete physical model (see Fig. 4.1). Equations (4.8) and (4.9) have a similar pattern. In both of them the associated elastic macrotensors **E** and **C** carry information about microstructure. But there is a crucial difference between **E** and **C**, and hence between the GCT and DM to be discussed.

Let us recall that in DM, solids with microstructure are studied on the basis of their discrete physical models. As soon as a certain physical model of a solid is chosen, all the information about the microstructure is assumed to be known. Otherwise it can be obtained by a realistic number of experiments (see, below, Sect. 4.5). By microstructure, we mean the two following groups of parameters of a physical model (see Fig. 4.2):

1. Physical parameters: microconstants of elastic interactions between constituent particles.
2a. Geometrical scale parameters: dimensions and separations of the constituent particles.
2b. Geometrical nonscale parameters: directions of the doublet axes.

Given this information, DM enables one to get the elastic macroconstants **C** as explicit functions of the above microstructural parameters. Specifically, for the physical model described by eqns. (4.9), the macroconstants **C** are expressed by (see eqn. (5.17))

$$C_{ijk_1\ldots k_h} = 2A_\circ \sum_{\alpha=1}^{n} \frac{\eta_\alpha^{h-2}}{h!} \overset{\circ}{\tau}_{\alpha i}\overset{\circ}{\tau}_{\alpha j}\overset{\circ}{\tau}_{\alpha k_1} \cdots \overset{\circ}{\tau}_{\alpha k_h}, \tag{4.10}$$

where $h = 2, 4, \ldots$. Equation (4.10) yields, in particular,

$$C_{ijkl} = A_\circ \sum_{\alpha=1}^{n} \overset{\circ}{\tau}_{\alpha i}\overset{\circ}{\tau}_{\alpha j}\overset{\circ}{\tau}_{\alpha k}\overset{\circ}{\tau}_{\alpha l}, \tag{4.11}$$

$$C_{ijklpq} = A_\circ \sum_{\alpha=1}^{n} \frac{\eta_\alpha^2}{12} \overset{\circ}{\tau}_{\alpha i}\overset{\circ}{\tau}_{\alpha j}\overset{\circ}{\tau}_{\alpha k}\overset{\circ}{\tau}_{\alpha l}\overset{\circ}{\tau}_{\alpha p}\overset{\circ}{\tau}_{\alpha q}, \tag{4.12}$$

$$C_{ijklpqrs} = A_\circ \sum_{\alpha=1}^{n} \frac{\eta_\alpha^4}{360} \overset{\circ}{\tau}_{\alpha i}\overset{\circ}{\tau}_{\alpha j}\overset{\circ}{\tau}_{\alpha k}\overset{\circ}{\tau}_{\alpha l}\overset{\circ}{\tau}_{\alpha p}\overset{\circ}{\tau}_{\alpha q}\overset{\circ}{\tau}_{\alpha r}\overset{\circ}{\tau}_{\alpha s} \tag{4.13}$$

These expressions include the following microstructural parameters of an underlying physical model:

1. A physical parameter $A_\circ$ which is a microconstant of elasticity (see Chap. 5).

2a. The geometrical scale parameters η_α (see Sect. 4.3.1).

2b. Geometrical non-scale parameters $\boldsymbol{\tau}_\alpha = \tau^\circ_{\alpha i}\mathbf{e}_i$ ($\alpha = 1, 2, \ldots, n$) which are the unit vectors of the doublets of α-directions (see eqn. (1.6)), $\mathbf{e}_i$ being the unit basis vectors of the fixed Cartesian frame of reference. The valence n depends on the underlying Bravais lattice and the kind of particle interactions. In the case of nearest neighbor interactions, the valence $n \leq 6$ regardless of the Bravais lattice.

Equations (4.10) through (4.13) show that, however high the rank of the macrotensors $\mathbf{C}$, they depend on a small number of the physical and geometrical microparameters. For instance, when the valence $n = 6$ we have in eqn. (4.10) only 13 microparameters: $A_\circ$, six η_α and six $\boldsymbol{\tau}_\alpha$. In general, the number N_{D} of the underlying microparameters in DM is limited to the value $N^*_{\mathrm{D}} = 2n + n_{\mathrm{p}}$, where n_{p} is the number of independent microconstants of elasticity which varies from 1, as in the above uniconstant version of DM, to a feasible maximum of 2 to 4. Thus, in the version of DM with nearest neighbor interactions, there may be no more than $N^*_{\mathrm{D}} = 2 \cdot 6 + 4 = 16$ microconstants. It must be noted that in practice the number N_{D} is significantly less than the maximum $N^*_{\mathrm{D}} = 16$. In particular,

- $N_{\mathrm{D}} = 2$ in the case of a uniconstant plane hexagonal lattice (see Chap. 5), which in a first (scaleless, continuum) approximation of DM is equivalent to an isotropic elastic continuum with Poisson's ratio $\nu = 1/4$.
- $N_{\mathrm{D}} = 3$ in the case of a biconstant plane simple cubic lattice (see Chap. 7), which in a first approximation of DM can be reduced to an isotropic elastic continuum with Poisson's ratio $\nu = 0$ with $N_{\mathrm{D}} = 2$.
- $N_{\mathrm{D}} = 4$ in the case of a triconstant simple cubic lattice, which in a first approximation of DM can be reduced to an isotropic elastic continuum with an arbitrary Poisson's ratio (see below Sect. 4.5.3) with $N_{\mathrm{D}} = 3$.

Once the microparameters of an underlying physical model in DM are specified, macroconstants of elasticity, such as C_{ijkl} and the like, can be computed by relations similar to eqn. (4.10).

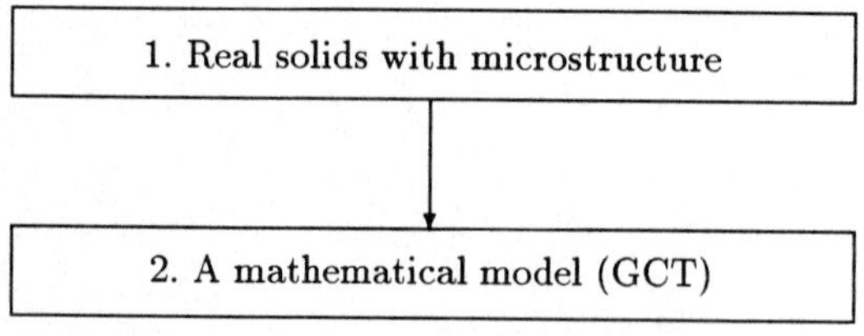

Fig. 4.3. Stages of developing the GCT

Quite the opposite situation exists in the GCT. To elaborate, we should point out that the GCT are phenomenological in nature. This means that unlike DM (see Fig. 4.1) they are developed following a simplified pattern (Fig. 4.3) without underlying physical prototypes. Under such an approach the parameters of the microstructure are not included in the mathematical model directly as in DM. The microstructural parameters enter into all of the GCT indirectly because they are *implicitly* contained in the macrotensors of elasticity E. In other words, unlike the elastic macrotensors C_{ijkl}, C_{ijklpq}, ... in DM, the elastic macrotensors E_{ijkl}, E_{ijklp}, ..., in the GCT are unknown functions of the underlying microstructural parameters. It follows that even though the microstructure of the solid is specified, the macrotensors E cannot—in contrast with the macrotensors C—be computed via the microstructural parameters. There remains only one way to determine the constants E and that is through experiments.

This way, however, is impeded by an immense number N_C of elastic constants E required. Indeed, while disregarding all kinds of symmetry relations peculiar to E, we can take that any additional subscript in the tensors $E_{ijk_1 \ldots k_h}$ triples the number of subsequent components $E_{ijk_1 \ldots k_{h+1}}$: there are 243 independent components E_{ijklp}, 729 components of E_{ijklpq}, and so on. Hence in eqn. (4.8) for materials of grade 2, which are the simplest microstructural generalization of classical continua, there are $N_C \approx 1000$ different elastic macroconstants E. The Mindlin theory (Mindlin 1964), representing the simplest microstructural generalization of Cosserat's continuum, has 903 independent elastic macroconstants. Both these numbers being close to 1000 are incomparably larger than that in DM which does not exceed the value $N_D^* = 16$. So even in the simplest GCT there are many hundreds of elastic macroconstants to be obtained experimentally. Clearly this is practically impossible, not to mention the additional macroconstants required in more complex theories of materials of grade three and higher.

Given microstructural symmetry, the number N_C is reduced. Nonetheless, it remains too large even at the highest-order symmetry. For instance, in Mindlin's *isotropic* microstructure (Mindlin 1964), $N_C = 18$ which is much larger than $N_D = 3$ for the above triconstant cubic lattice reduced to an *isotropic* one in a continuum approximation of DM. For comparison, we can refer to the complexity of experiments described by Lakes (1995) to obtain only six Cosserat constants (not Mindlin's eighteen) using the theoretical results for a bending of rods (Krishna Reddy and Venkatasubramanian 1978) and plates (Gauthier and Jahsman 1975).

We should also point out an additional salient feature of the GCT. Suppose for a moment that we are able to get all the elastic macroconstants E in the theories of materials of grades $N = 1, 2, 3$, etc. Then applying these theories sequentially, we would be monotonously approaching an exact description of the elastic behavior of the solid in question. Since such a promising process is impossible, and since we are forced to stop at materials of grade 2 or even 1, the theoretical description of elasticity by a GCT remains approximate. Any abridged GCT represents an approximate theory of a real solid or an exact theory of its approximate image. By analogy with DM (see Fig. 4.1) we can thus extend the pattern in Fig. 4.3 and depict it as in Fig.4.4. Although the second element of the pattern—an approximate image of a solid in Fig. 4.4—resembles its counterpart in Fig. 4.1, a discrete physical model, they are in fact drastically different from each other. This difference is important and is, therefore, to be elucidated.

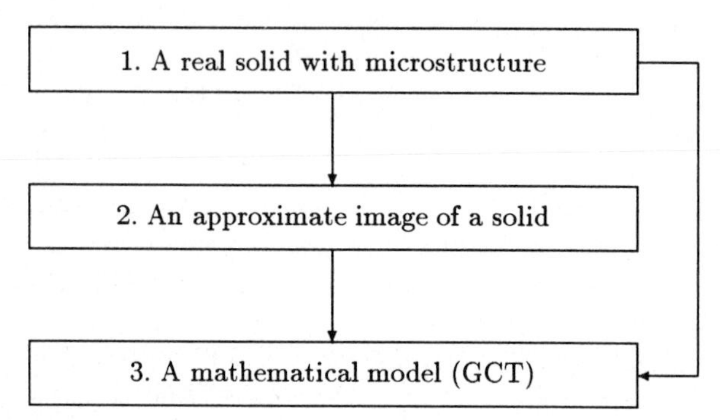

Fig. 4.4. Stages of developing the GCT (an extended pattern)

A discrete physical model in Fig. 4.1 is chosen in accord with a real solid, *before* a mathematical model. By an appropriate choice of the physical model, one is able to draw it closer to a real solid and thus decrease EMM-1 (see Sect. 4.3.1) and increase the exactness of DM. Unlike the physical model of DM, an approximate image of a solid in Fig. 4.4 cannot be chosen first—it is generated automatically *after* the GCT is selected. Since the only reasonable GCT can be that of materials of grade $N = 2$, the gap between a solid and its approximate image is frozen and cannot be narrowed by going to the other more accurate GCT of materials of grade $N > 2$. The GCT are, therefore, not amenable to refinement.

The above discussion of the GCT may be summarized as follows:

1. The number of elastic macrotensors N_C in any GCT is, in general, extremely large and they cannot be obtained by means of a realistic number of experiments.
2. Due to the above disadvantage, the GCT are, in general, inapplicable as a result of the unrealistic number of experiments that would be required.
3. The GCT can be applied only to materials of grade 2 and only in some particular cases when the above number N_C can be reduced by considering relatively simple problems (e.g., plane waves) for solids of simplified forms (rods, plates, etc.) and isotropic symmetry of microstructure.
4. Unlike DM, the GCT are not amenable to refinement.

4.4.3 Nonlocal Theories of Integral Type

General Remarks. The integral nonlocal theories (No. 4 in Table 4.1) are based on the assumption that the forces of interaction between atoms in a solid extend beyond nearest neighbors. They have a characteristic radius of action $l^* > \eta$, where η is the nearest neighbor separation. In general, the radius l^* is unbounded. However, outside a certain limit l the long-range interatomic forces drop considerably and may be neglected in accord with a problem under study. The limit l depends on the type of constituent atoms and on the character of the problem in question. It also depends on the underlying model of the interatomic forces (UMIF).

Information about the interatomic forces is obtained primarily by neutron spectroscopy (Bacon 1962) which allows one to measure phonon frequencies. The experimental data are then correlated with the theoretical frequencies delivered by various UMIF that include different ranges of action $R \equiv l/\eta$. Such a method was applied, in particular, to covalently bonded crystals of germanium (Cochran 1973). It has been determined that a good fit to the experimental frequencies can be achieved by a "rigid ion" model[7] which takes into account the interatomic forces extending up to the fifth-nearest neighbors, i.e., having the range $R = 5$. At the same time a more sophisticated "shell" model[8] has given quite good agreement with the measurements provided that only nearest shells and cores interact, i.e., when $R = 1$. As a matter of fact, the range R may reach 15 (Cochran 1963), which is probably an upper bound for any interatomic forces.

Due to the long-range interatomic forces, integral type theories involve the scale parameter l and thus assume nonlocal character. The integral nonlocal theories (INT) have been pioneered by Kröner (1967), Krumhansl (1968), Mindlin (1968), Edelen et al (1971), and Eringen and Edelen (1972, 1973). A comprehensive exposition of the INT is presented in the work by Edelen (1976) and in the monographs by Kunin (1982, 1983). We present here a sketch of the INT that is suitable for a comparison of the INT with other nonlocal theories of elasticity, viz., DM and the GCT.

Nonlocal Equations of Motion. We begin with the simplest one dimensional discrete model of a crystal. It is an infinite homogeneous linear chain of identical atoms of mass m located along the x-axis. The n-th atom has an equilibrium coordinate $x_n = n\eta$ where η is the nearest atom separation and n belongs to the set $\mathbb{Z}$ of all integers. The masses are bonded to each other by interatomic elastic forces having the range of action from 1 to R.

[7] A central force adiabatic model of LD for ionic crystals in which rigid and unpolarizable ions interact by Coulomb's long-range cohesive forces and short- range repulsive forces with $R = 1$ (Maradudin et al 1971, pp. 221–225).

[8] It includes both the ion-ion and the ion-electron interactions: each ion is regarded as a core to which a massless shell, representing the outer valence electrons, is bound by short-range isotropic forces (Böttger 1983, pp. 59–60).

Displacements of the atoms $u(n,t)$ at any instant of time t are assumed to be longitudinal, i.e., $u(n,t) \parallel x$.

The equation of motion of the chain in the particular case with the maximum range of action $R = 1$ is

$$m\ddot{u}(n) = F(n,1) + P(n). \tag{4.14}$$

The function $F(n,1)$ is given by

$$F(n,1) \equiv \alpha(1)[u(n+1) - u(n)] + \alpha(-1)[u(n-1) - u(n)] \tag{4.15}$$

where $\alpha(1)$ and $\alpha(-1)$ are the force constants (or, simply, the stiffness coefficients) of the elastic interatomic bonds having the range of action $r = R = 1$ and $P(n)$ is an external force acting on the atom n. In the above notation we have, for simplicity, omitted the argument t from the time dependent functions $u(n,t)$ and $P(n,t)$.

In the case of $R = 2$, eqn. (4.14) will include the additional terms $F(n,2)$ representing the influence of the second range interatomic bonds characterized by the force constants $\alpha(2)$ and $\alpha(-2)$:

$$m\ddot{u}(n) = F(n,1) + F(n,2) + P(n). \tag{4.16}$$

The function $F(n,2)$ is analogous to $F(n,1)$:

$$F(n,2) \equiv \alpha(2)[u(n+2) - u(n)] + \alpha(-2)[u(n-2) - u(n)] \tag{4.17}$$

Continuing this process we easily arrive at the equation of motion of the chain in the general case of arbitrary R

$$m\ddot{u}(n) = \sum_{r=-R}^{R} \alpha(r)[u(n+r) - u(n)] + P(n). \tag{4.18}$$

The stiffness coefficients $\alpha(r)$ in eqn. (4.18) hold for the elastic interatomic bonds whose range of action is $|r|$. Due to the homogeneity of the chain, the constants $\alpha(r)$ are even functions of r, i.e.,

$$\alpha(r) = \alpha(-r) = \alpha(|r|) \tag{4.19}$$

where

$$r \in \mathbb{Z}_R = \{-R, -R+1, \ldots, R-1, R\} \subseteq \mathbb{Z}, \tag{4.20}$$

where K is the set of integers. Note that in eqn. (4.18) the constant $\alpha(0)$ is multiplied by zero and may therefore assume any finite value; we take $\alpha(0) = 0$.

In view of eqn. (4.19), eqn. (4.18) can also be presented as the sum of symmetric terms

$$m\ddot{u}(n) = \sum_{r=1}^{R} \alpha(r)[u(n+r) - 2u(n) + u(n-r)] + P(n). \tag{4.21}$$

If we now take $n' \equiv (n+r) \in \mathbb{Z}$ and introduce new force constants $\Phi(n-n')$ so that

$$\Phi(n-n') \equiv \Phi(r) = -\alpha(r), \qquad r \neq 0, \tag{4.22}$$

$$\Phi(0) = \sum_{r=-R}^{R} \alpha(r), \tag{4.23}$$

then eqn. (4.18) yields

$$m\ddot{u}(n) + \sum_{n'=-R}^{n+R} \Phi(n-n')\, u(n') = P(n). \tag{4.24}$$

The relations (4.19) and (4.22) yield

$$\Phi(n-n') = \Phi(n'-n) = \Phi(|n-n'|). \tag{4.25}$$

As is clear from eqns. (4.22) and (4.23), the new force constants $\Phi(n-n')$ are different from zero only on $\mathbb{Z}_R$. Beyond the set $\mathbb{Z}_R$, the parameters $\Phi(n-n') = 0$. In other words,

$$\Phi(n-n') \begin{cases} \neq 0, & |n-n'| \leq R \\ = 0, & |n-n'| > R. \end{cases} \tag{4.26}$$

Owing to eqn. (4.26), eqn. (4.24) can be rewritten in the form

$$m\ddot{u}(n) + \sum_{n'} \Phi(n-n')\, u(n') = P(n) \tag{4.27}$$

which coincides with Kunin (1982, eqn. 2.1.13) where this relation was obtained in a different way. Following the conventional procedure , , (Askar 1985, p. 64, Maradudin 1971, p. 576), we make the transformations $m \mapsto \rho$, $n \mapsto x \in \mathbb{R}$, $n' \mapsto x' \in \mathbb{R}$, $P(n) \Rightarrow q(x)$, where ρ is the mass density, $q(x)$ is the external force distribution, and then replace summation over n' by integration over x' according to

$$\sum_{n'} F(n,n') \mapsto \int_{-l}^{l} F(x,x')\, dx'. \tag{4.28}$$

On the basis of eqn. (4.28) and with the argument t restored, eqn. (4.27) transforms into

$$\rho\ddot{u}(x,t) + \int_{-l}^{l} \Phi(x-x')\, u(x',t)\, dx' = q(x,t). \tag{4.29}$$

Equation (4.29) is the same as (Kunin 1982, eqn. 2.4.39) and represents the key relation in a one-dimensional INT. In accord with eqn. (4.26), the kernel $\Phi(x-x')$ in eqn. (4.29) must satisfy the condition of finiteness

$$\Phi(x-x') \begin{cases} \neq 0, & |x-x'| \leq l \\ = 0, & |x-x'| > l \end{cases} \tag{4.30}$$

Let us recall that the difference equation (4.27) relates to an infinite homogeneous chain of identical atoms. Therefore the integral equation (4.29) derived from (4.27) is valid, in general, for only infinite one-dimensional solids. At the same time, the conditions (4.30) reduce this limitation in that they make eqn. (4.29) applicable also to finite and semi-infinite solids with the exception of their boundary regions (layers) of length l.

Boundary Problems. Since any theory of elasticity is primarily intended for solving boundary problems, so are—or at least should be—all the INT. Therefore many intrinsic features of the INT can be revealed by considering the way they deal with boundary problems. Here we touch upon this matter in terms of a one-dimensional elastic solid occupying the semi-infinite domain $-l \leq x < \infty$. The segment $-l \leq x \leq 0$ and the semi-axis $0 < x < \infty$ correspond to the boundary layer S and the interior D of the solid, respectively. Restricting ourselves to the static case, we rewrite eqn. (4.29) for the region D as follows

$$\int_{-l}^{\infty} \Phi(x - x')\, u(x')\, dx' = q_D(x), \qquad x \in D \tag{4.31}$$

where the kernel $\Phi(x - x')$ obeys eqn. (4.30). Equation (4.31) should be supplemented by boundary conditions. As in classical elasticity, these are of the two basic forms (Kunin 1982):

$$\int_{-l}^{l} \Gamma(x, x')\, u(x')\, dx' = q_S(x), \qquad x \in S \tag{4.32}$$

$$u_S(x) = h(x), \qquad x \in S \tag{4.33}$$

which apply to the first and the second boundary problems, respectively. In the above relations, $q_S(x)$ and $h(x)$ are external forces and displacements to be specified on S. The notation of Kunin (1982) has been adopted. The basic boundary equations (4.31) through (4.33) have been analyzed in detail in literature on the INT. Some remarkable features of these relations are briefly considered below.

We begin with the second boundary problem. To solve it, we must adopt a certain distribution of the displacements $u_S(x)$, i.e., specify the function $h(x)$ on the segment $S = [-l, 0]$. This is quite a new and difficult requirement that has never emerged in the other theories of elasticity. Its difficulty becomes clear while we draw on the analogy between the above basic boundary problems. In the first one instead of the displacements $u_S(x)$, the external forces $q_S(x)$ are to be given on S. This can be done in the usual way by specifying $q_S(x) = q \neq 0$ only on the surface of a solid, i.e., at the point $x = -l$. At the other points $x \in (-l, 0]$ of the boundary layer S the forces $q_S(x)$ can be taken equal to zero as assumed in all the theories of elasticity except the INT:

$$q_S(x) = \begin{cases} q, & x = -l \\ 0, & x \neq -l \end{cases} \tag{4.34}$$

Let us now suppose for a moment that the boundary displacements $u_S(x)$ are distributed on S similarly, i.e.,

$$u_S(x) = h(x) = \begin{cases} h, & x = -l \\ 0, & x \neq -l \end{cases} \tag{4.35}$$

It is easy to see that the distribution (4.35) is preposterous because it describes the decomposition of the boundary layer S by peeling its surface $x = -l$ away or pushing it through S. To avoid this senselessness, the function $h(x)$ on S must be continuous and this is the only one point known for sure. But there are innumerable continuous functions and no guides on how to select a unique $h(x)$ among them *in advance.* Of course one may consider any continuous function $h(x)$ in the abstract. However, in the most realistic cases of the second boundary problem—in the contact problems of elasticity (Gladwell 1980, Kikuchi and Oden 1988)—an arbitrary choice of $h(x)$ is inadmissible. The point is that in a contact problem, the boundary displacements $h(x)$ are given at the surface $x = -l$, i.e., on a part S_0 of the boundary layer S. On the rest of the region S, at $x \in (S - S_0) = (-l, 0]$ the function $h(x)$ is unknown *beforehand* and can only be found *after* a contact problem is solved.

Let us now touch upon the first basic boundary problem described by the integral equations (4.31) and (4.32) which include quite distinct kernels—difference $\Phi(x-x')$ and nondifference $\Gamma(x, x')$. Being nondifference, the kernel $\Gamma(x, x')$ entails a significant complication of the problem (Kunin 1982, p. 70):

> ...in the general case the first basic problem is equivalent to the solution of an infinite system of linear algebraic equations. Therefore a solution in the closed form can be obtained only in the approximation by the first roots.

These algebraic equations remain infinite at any parameters l, either large or small, if only $l > \eta$. Meanwhile, as noted above, the interatomic forces do not span appreciably beyond approximately 15 interatomic spacings η so that at worst $l = 15\eta$. More feasible values of l fall within the range between two and five distances η. In these cases, the first boundary problem can be efficiently treated in terms of the difference equations (4.18) or (4.27) from which the INT are derived. Such an approach does not give rise to the foregoing infinite algebraic equations and admits not only numerical but also analytical solutions in the closed form (Granik and Ferrari 1996).

Within the framework of the INT, the one-dimensional boundary problems have the following features:

1. The first boundary problem can be easily formulated but its solution runs into serious mathematical difficulties. In many cases, the difficulties may be obviated by applying the simpler difference equations (4.18) or (4.27).
2. The second boundary problem can sometimes be solved with lesser mathematical difficulties but its solution is often impeded by the uncertainty of the boundary condition (4.33).

In the cases of two- and three-dimensional boundary problems, these difficulties considerably increase and along with some additional theoretical obstacles make the INT, in general, intractable. As a result, the applications of the INT have until now been limited to several elementary one-dimensional *boundary problems* and some two- and three-dimensional *non-boundary problems* for which lattice dynamics is also applicable.

4.5 Doublet Mechanics: Closure

4.5.1 Advantages

The foregoing discussion has shown several limitations of comparable theories which are not inherent in DM and give it certain advantages. These advantages allow us to delineate some domains of efficient applicability of DM. Before doing this it is useful to make a summary of the theories considered above in comparison with DM.

Lattice Dynamics

- Lattice dynamics is a micromechanics of infinite elastic crystals and is, in general, unable to deal with boundary problems which are the main concern of all the theories of elasticity including DM.

Classical Theory of Elasticity

- Unlike DM, classical elasticity disregards the discrete nature of solids and is therefore inapplicable for studying scaling microstructural phenomena.

Cosserat's Theory of Elasticity

- If the microstructure is nonaxial, then a first approximation of DM coincides with Cosserat's elasticity. Such an approximation is often insufficient to describe scaling effects fairly well and can be improved by the second and higher approximations of DM that are completely different from the Cosserat theory.
- If the microstructure is axial, then Cosserat's theory reduces to the scaleless classical elasticity and therefore, in contrast to DM, fails to represent scale phenomena.

Nonlocal Micromechanical Theories of Differential Type

- These theories take into account the microstructure of elastic solids and should, in principle, be able to study scale phenomena. Unlike DM, however, they include a great number of phenomenological macroconstants that cannot be determined by a realistic number of experiments. Due to this obstacle, differential type theories are impractical except some simple cases.

Nonlocal Micromechanical Theories of Integral Type

- In contrast to all other models of elasticity, the integral type micromechanical theories are based on the assumption of long-range elastic interactions of constituent particles. However, an overcomplicated mathematical technique and the uncertainty of the displacement boundary conditions make these theories either needlessly difficult (in the case of one-dimensional boundary problems) or even generally intractable (in the cases of two- and three- dimensional boundary problems).

The Theories of Elasticity as a Whole

- All the theories of elasticity, except DM, deal with various tensors of stresses. The stress tensors make real sense only with respect to continua in which the action of a force applied at any internal point is transmitted in all directions. On the other hand, the action of a force applied at any material particle of a solid with discrete microstructure is transmitted only in specified directions, from one constituent particle to another. This fundamental feature of discrete microstructure cannot be described within the framework of the tensor-based theories of elasticity.
- In contrast to these theories DM employs a technique of doublet microstresses that, being either scalars or vectors, adequately represent microforces and microcouples of the *directional* particle interactions. In comparison with the stress tensors, the doublet microstresses constitute a more precise tool which imparts to DM a deeper insight into the mechanical behavior of solids with microstructure and thus makes it possible to discover new micromechanical phenomena.

4.5.2 Domains of Efficient Applicability

Below we outline some domains of efficient applicability of DM. We do not, however, specify the microstructural solids for which they apply. This specification is considered later, in Sect. 4.5.3. Thus we have

In any approximation of DM

1. Development of new micromechanical versions of the following theories
   1.1. Viscoelasticity
   1.2. Plasticity

In a first (local) approximation of DM

If the number of doublet microstresses N_{DM} is larger than that of macrostresses N_{TM} ($N_{\mathrm{DM}} > N_{\mathrm{TM}}$):

2. New solutions of boundary problems in terms of doublet microstresses

- 2.1. Dynamic boundary problems
  - 2.1.1. Vibrations of soil bases (foundations)
  - 2.1.2. Propagation of surface waves
- 2.2. Static boundary problems
  - 2.2.1. Strains and stresses in soil bases (foundations)

3. New solutions of stress concentration problems. Determination of doublet microstresses
   - 3.1. Around holes, cavities, notches, and the like
   - 3.2. Around inclusions
   - 3.3. Under punches, beams, shells, etc.
   - 3.4. At the tips of cracks
4. Transfer from doublet microstresses to conventional macrostresses
   - 4.1. Studies of macrostresses unobtainable by the tensor-based theories
   - 4.2. Computations of macrotensors of elasticity via parameters of microstructure
   - 4.3. Solutions of homogenization problems

If $N_{\mathrm{DM}} = N_{\mathrm{TM}}$:

5. If the problems #2 and 3 have already been solved on the basis of macrostresses, then
   - 5.1. Transfer from macrostresses to doublet microstresses
   - 5.2. Studies of scaleless microstructural phenomena in terms of doublet microstresses
6. If the problems #2 and 3 have not been solved on the basis of macrostresses, then
   - 6.1. Solutions of the problems #2 and 3

If $N_{\mathrm{DM}} \geq N_{\mathrm{TM}}$:

7. Development of new failure criteria on the basis of doublet microstresses.

In nonlocal approximations of DM

8. Development of new failure criteria including not only the doublet microstresses but also their spatial gradients and scale parameters of microstructure.
9. Solutions of the boundary problems #2 in view of scale parameters of microstructure.
10. Solutions of the stress concentration problems #3 in view of scale parameters of microstructure.
11. Development of a new version of fracture mechanics using solutions of the problems #10 (here, the stress concentration factor is a *finite* quantity, in contrast with all the local theories of elasticity).

4.5.3 Real Solids, Physical Models and Microstructural Parameters

In Sect. 4.4.2 we considered one of the most influential features of DM: a small number N_D of microstructural parameters involved in any approximation of this theory. In the case of nearest neighbor interactions, the number $N_\mathrm{D} \leq N^*_\mathrm{D} = 16$ is even less than the maximum number of macroconstants $N^*_\mathrm{C} = 21$ in classical anisotropic elasticity (CAE). Just as the number N^*_C in CAE decreases owing to various kinds of symmetry, so does N^*_D in DM. As a result, in most realistic cases of solids with microstructure, the number of microparameters of DM varies over a narrow range $2 \leq N_\mathrm{D} \leq N^*_\mathrm{D} = 5$[9]. Determination of these microparameters is a vital part of DM and deserves a special consideration. In this chapter we outline conceptual facets of this momentous problem.

A solution of the problem starts from choosing a discrete physical model of DM that has to represent a real solid (see Fig. 4.1). This step is not unique. It is intrinsic to all microstructural theories including in particular LD. Referring to LD makes it easier to understand the way of transferring from real solids to their physical models.

For lattice dynamics, the real solids under study are crystals, objects of great complexity to tackle without making decisive simplifications. Lattice dynamics adopts them by introducing tractable physical models of real crystals. Different models make different compromises that is why there are a variety of such models in LD, from the simplest Born-von Kármán elastic chain of atoms with short-range interatomic forces (see eqns. (4.14) and (4.15)) to more sophisticated models such as a "rigid ion" model with short-range repulsive and long-range attractive microforces, a "shell" model (see Sect. 4.4.3), a "breathing shell" model (Madelung 1978), and so on.

These examples clearly show that the choice of a suitable physical model is not a matter of philosophy—it is a matter of experimental test. And as such, it is dictated first and foremost by the obvious test requirements:

1. Parameters of a physical model have to be amenable to either direct or indirect experimental determination.
2. A mathematical theory based on a physical model has to be capable of describing experimental phenomena.
3. The theory should be able to predict new phenomena supported then by new experiments.

Let us now return to DM. The basic physical model of DM (see Chap. 1 and Sect. 4.3.1) is not a "frozen" object—it admits a variety of particular variants with different geometrical and physical microstructural parameters. This variability enables DM to deal with a number of real solids possessing microstructure of different scale, from crystals to granular packings and arrays

[9] The same range in CAE includes elastic solids possessing isotropic ($N_\mathrm{C} = 2$), cubic ($N_\mathrm{C} = 3$), and transversely isotropic ($N_\mathrm{C} = 5$) symmetry.

of tectonic plates in the Earth's crust. Each of these solids can be represented by a variant of the basic physical model of DM. Whether the variant is suitable or unsuitable for a given solid ought not to be decided *a priori*— this question can be resolved only *a posteriori*, in terms of the above test requirements. Before this step is taken, the physical model chosen should not be ruled out. In general, only experience will exactly determine the kinds of solids and the scope of problems that are analyzable by DM.

We can draw an important conclusion: since classical elasticity and Cosserat's theory are particular cases of DM, all the solids and problems they study can also be treated by DM. An example will make it clear.

Let us take a variant of an elastic isotropic continuum characterized by the Lamé macroconstants λ and μ. Within the framework of classical elasticity, this continuum is a reasonable physical model of certain real solids. On the other hand, as shown below, a discrete physical model of DM with a simple cubic microstructure described by three independent elastic microconstants: $A_\circ$, $B_\circ$ and $I_\circ$, and a scale parameter η (the interparticle separation) can be reduced in a first approximation of DM to an isotropic continuum with only two independent microconstants $A_\circ = \lambda + 2\mu$, $I_\circ = 2\mu$, $B_\circ = A_\circ - I_\circ = \lambda$, the scale parameter η being equal to zero. It follows that for the same real solids, this particular physical model of DM is no less reasonable and no doubt more general than the classical isotropic continuum. We thus can transfer from classical elasticity to DM and consider all the isotropic scaleless continua as a particular case of discrete solids that possess scaling cubic microstructure with the elastic microconstants $A_\circ = \lambda + 2\mu$, $I_\circ = 2\mu$, $B_\circ = \lambda$ and the scale parameter $\eta > 0$.

There might be an objection that real isotropic media are those of chaotic microstructure which is far from any kind of cubic symmetry. But this objection is unessential because, in comparison with classical isotropy, cubic symmetry leads to the same *macrostresses* in a continuum approximation of DM. Moreover, in the same approximation, DM can render additional important information: *microstresses*. This information is inaccessible to classical elasticity and alone gives DM a significant advantage. Furthermore, if we apply DM in scale $M > 1$ approximations then the influence of the scale microparameter η on the elastic behavior of a solid in question will be revealed. If the cubic (or any other physical) model satisfies the above test requirements then its application to any elastic solid treated by classical and Cosserat's elasticity is also completely valid and, in addition, much more informative.

Classical elasticity is applicable to crystals when modeled as anisotropic continua. Therefore DM being a scaling generalization of classical elasticity has at least adequate reason for applying to crystals in view of their discrete nature. This is, however, possible under some limitations: (i) the crystal structure should be a Bravais lattice and (ii) the microforces of interaction between atoms have to be short-range or moderate long-range extending respectively to nearest or next-nearest and third-nearest neighbors.

Despite certain deficiencies (Maradudin et al 1971, pp. 68–77), crystal models including a small number of force constants, from one to three, continue to be an efficient tool for studying lattice vibrations. In particular, on the basis of a simple cubic lattice with one force constant, a theory of surface phonons in superlattices was developed (Dobrzynski et al 1984). Even the simplest Born-von Kármán chain model with nearest-neighbor interactions is used efficiently to study rather subtle physical phenomena in crystals, e.g., zone-folding effects (Barker, Jr. et al 1978). Since LD is, in general, inapplicable to boundary problems of elasticity, DM, in the version with up to third nearest neighbor interactions, can be successfully applied to these problems for certain crystal structures. Two such applications are presented in the article (Granik and Ferrari 1996) and in Chap. 7 of this book.

There is also a special kind of microstructural solids, the so-called granular media, whose constituent particles interact only through nearest neighbor contact microforces. Granular solids display both elastic and inelastic properties. The latter may often be neglected. In this case, granular media can be considered as elastic solids with microstructure and efficiently treated by DM (Granik and Ferrari 1993).

It should be noted that for many decades elastic granular solids have been the subject of inquiry in numerous theoretical works carried out on the basis of discrete and continuum approaches. A comparison of DM with these works is of interest and deserves a detailed analysis. This analysis, however, would extend beyond the scope of this chapter. A brief discussion of this matter can be found in Chap. 5 and in the article by Granik and Ferrari (1993).

In connection with granular media, it is worth emphasizing one more advantageous feature of DM. It stems from the fact that DM is developed as a general theory covering different microstructural solids under any mechanical actions rather than an *ad hoc* theory for some particular solids at some particular mechanical actions. It follows that DM can provide more information than any comparable *ad hoc* model. As an illustration, let us take a look at a boundary problem for a granular semispace loaded by a vertical concentrated surface force. A solution of the problem by means of DM is given in Chap. 8. The solution has revealed a "conical" transmission of a surface load into the depth of the granular array. This phenomenon is well-known from experiments but is completely concealed from all the tensor-based theories of elasticity. That is why several *ad hoc* models have been developed that are capable of describing this specific phenomenon (Misra 1979, Cundal and Strack 1979, Li and Bagster 1990, 1993, Liffman et al. 1992, Beranek and Hobbelman 1992, Meek and Wolf 1993). Unlike DM, the *ad hoc* models are unable to represent other essential phenomena in granular media such as Rayleigh waves.

The capability of studying wave propagation in discrete solids can make DM a promising tool in geophysics and seismology. The point is that due to faults, the Earth's crust is divided into numerous tectonic blocks which are on the order of 15 to 25 km in length and depth (Robinson and Benites 1995). There are other discontinuities such as stress-induced cracks, preferentially oriented pores and impurities (Anderson et al. 1974, Garbin and Knopoff 1975, Hudson 1982, Crampin 1981, 1984, Fraser 1990). Due to these discontinuities, the Earth's crust can be considered as a solid with a pronounced stochastic microstructure and as such can be dealt with by DM. Doublet Mechanics can be especially useful in a study of waves of short wavelength. These waves being most sensitive to the particulate structure of the Earth's crust exhibit some phenomena (e.g., strong scattering effects) that so far have not been completely accounted for by continuum scaleless theories (Pollitz 1994).

Let us now turn to microscopic parameters of DM. We suppose that for a given real solid, a particular variant of the basic physical model of DM is chosen and hence the kind and the number of microparameters are fixed. If unknown, they have to be determined by using experimental information. Here we will exemplify a possible approach to a solution of the problem. As a physical model we select a simple cubic microstructure with independent translations $\mathbf{u}$ and rotations $\boldsymbol{\phi}$ of constituent particles. Further analysis is based on the governing equations of DM presented in Sect. 1.4 (see also Sect. 4.3). In order to emphasize conceptual facets of the problem, we omit insignificant details.

Neglecting doublet microstresses of torsion ($\mathbf{m}_\alpha \equiv 0$), we consider a non-polar microstructure in which only doublet microstresses of elongation $\mathbf{p}_\alpha$ and shear $\mathbf{t}_\alpha$ remain ($\alpha = 1, 2, 3$). For these microstresses, we take a variant of constitutive *microequations* with three independent microconstants of elasticity $A_\circ$, $B_\circ$ and $I_\circ$:

$$\begin{Bmatrix} \mathbf{p}_1 \\ \mathbf{p}_2 \\ \mathbf{p}_3 \end{Bmatrix} = \begin{bmatrix} A_\circ & B_\circ & B_\circ \\ B_\circ & A_\circ & B_\circ \\ B_\circ & B_\circ & A_\circ \end{bmatrix} \begin{Bmatrix} \epsilon_1 \\ \epsilon_2 \\ \epsilon_3 \end{Bmatrix} \tag{4.36}$$

$$\begin{Bmatrix} \mathbf{t}_1 \\ \mathbf{t}_2 \\ \mathbf{t}_3 \end{Bmatrix} = \begin{bmatrix} I_\circ & 0 & 0 \\ 0 & I_\circ & 0 \\ 0 & 0 & I_\circ \end{bmatrix} \begin{Bmatrix} \gamma_1 \\ \gamma_2 \\ \gamma_3 \end{Bmatrix} \tag{4.37}$$

where $\boldsymbol{\epsilon}$ and $\boldsymbol{\gamma}$ are doublet microstrains of elongation and shear, respectively. It is easy to show that in a first (scaleless, continuum) approximation of DM, eqns. (4.36) and (4.37) along with the macro-microrelations (see eqns. (1.40) and (1.41)) bring about the following constitutive *macroequations* relating macrostresses σ_{ij} to macrostrains ε_{ij}:

$$\begin{Bmatrix} \sigma_{11} \\ \sigma_{22} \\ \sigma_{33} \\ \sigma_{12} \\ \sigma_{23} \\ \sigma_{31} \end{Bmatrix} = \begin{bmatrix} A_\circ & B_\circ & B_\circ & 0 & 0 & 0 \\ B_\circ & A_\circ & B_\circ & 0 & 0 & 0 \\ B_\circ & B_\circ & A_\circ & 0 & 0 & 0 \\ 0 & 0 & 0 & I_\circ & 0 & 0 \\ 0 & 0 & 0 & 0 & I_\circ & 0 \\ 0 & 0 & 0 & 0 & 0 & I_\circ \end{bmatrix} \begin{Bmatrix} \varepsilon_{11} \\ \varepsilon_{22} \\ \varepsilon_{33} \\ \varepsilon_{12} \\ \varepsilon_{23} \\ \varepsilon_{31} \end{Bmatrix} \tag{4.38}$$

Equations (4.38) characterize a linear elastic cubic continuum. We thus see that the elastic *microconstants* $A_\circ$, $B_\circ$ and $I_\circ$ are at the same time elastic *macroconstants*. This fact greatly facilitates the problem because macroconstants can be determined without special subtleties peculiar to microtests, just in terms of rather "rough" and simple macroscopic experiments. On the other hand, if the solid in question is a cubic crystal, we can simply borrow the values of $A_\circ$, $B_\circ$ and $I_\circ$ from experimental crystal data. In particular, for a cubic crystal of NaCl we have (in 10^{11} dyne/cm^2) $A_\circ \equiv c_{11} = 5.73$, $B_\circ \equiv c_{12} = 1.12$ and $I_\circ \equiv c_{44} = 1.33$ (Catlow et al 1982, p. 147, Table 3). In the case of isotropy, eqns. (4.38) are transformed into (Gould 1983)

$$\begin{Bmatrix} \sigma_{11} \\ \sigma_{22} \\ \sigma_{33} \\ \sigma_{12} \\ \sigma_{23} \\ \sigma_{31} \end{Bmatrix} = \begin{bmatrix} 2\mu+\lambda & \lambda & \lambda & 0 & 0 & 0 \\ \lambda & 2\mu+\lambda & \lambda & 0 & 0 & 0 \\ \lambda & \lambda & 2\mu+\lambda & 0 & 0 & 0 \\ 0 & 0 & 0 & 2\mu & 0 & 0 \\ 0 & 0 & 0 & 0 & 2\mu & 0 \\ 0 & 0 & 0 & 0 & 0 & 2\mu \end{bmatrix} \begin{Bmatrix} \varepsilon_{11} \\ \varepsilon_{22} \\ \varepsilon_{33} \\ \varepsilon_{12} \\ \varepsilon_{23} \\ \varepsilon_{31} \end{Bmatrix} \tag{4.39}$$

where μ and λ are the Lamé constants. The comparison of eqns. (4.38) with eqns. (4.39) yields

$$A_\circ = 2\mu + \lambda, \qquad I_\circ = 2\mu, \qquad B_\circ = A_\circ - I_\circ = \lambda \tag{4.40}$$

The relations (4.40) reduce the problem of determining $A_\circ$, $B_\circ$ and $I_\circ$ to the problem of getting the Lamé constants. The latter problem presents no special difficulties.

Besides the *physical* parameters $A_\circ$, $B_\circ$ and $I_\circ$, the cubic microstructure is characterized by the *geometrical scale* parameter η, the interparticle separation. If the real solid in question is a crystal then η can be found similarly to $A_\circ$, $B_\circ$ and $I_\circ$ from experimental crystal data. For example, $\eta = 2.82$Å for a crystal of NaCl (Catlow et al 1982, p. 147, Table 3). In all other cases, when a microstructural solid is not a crystal, the scale parameter η can be obtained in terms of dispersion relations. Let us touch upon this method.

For simplicity we consider the case of one-dimensional deformations of the elastic cubic structure. In this case, the governing equations of DM can easily be reduced to the difference equation (4.21) of longitudinal vibrations in which the range of action $R = 1$ (for nearest neighbor interactions) and $\alpha(1) = A_\circ\eta$. Assuming a longitudinal coordinate $x_n = n\eta$ and taking $u(x_n) \equiv u_n$, $P(n) = 0$, we bring eqn. (4.21) to the form

$$m\ddot{u}_n = A_\circ\eta(u_{n+1} - 2u_n + u_{n-1}). \tag{4.41}$$

Equation (4.41) admits a solution of the type of a plane steady-state wave propagating along the x-axis:

$$u_n = U\exp[i(kx_n - \omega t)] \tag{4.42}$$

where U, k and ω are the wave amplitude, the wavenumber and the angular frequency, respectively. Inserting u_n from eqn. (4.42) into eqn. (4.41) and taking into account that $m = \rho\eta^3$, $x_{n\pm 1} = x_n \pm \eta$, we come to the dispersion relation

$$\omega = \frac{2}{\eta}\sqrt{\frac{A_\circ}{\rho}}\sin(k\eta/2). \tag{4.43}$$

This relation also allows one to obtain the phase velocity $V_\mathrm{p} \equiv \omega/k$. In view of the identity $k \equiv 2\pi/\lambda$, λ being the wavelength, eqn. (4.43) yields

$$V_\mathrm{p} \equiv \omega/k = V_\circ \frac{\sin(\pi\eta/\lambda)}{\pi\eta/\lambda} \tag{4.44}$$

where $V_\circ \equiv \sqrt{A_\circ/\rho}$ is the velocity of sound, i.e., the velocity of the wave when the wavelength $\lambda \to \infty$ or the scale parameter $\eta \to 0$. As seen from eqns. (4.43) and (4.44), the frequency ω and the phase velocity V_p depend on the scale parameter η: $\omega = \omega(\eta)$, $V_\mathrm{p} = V_\mathrm{p}(\eta)$. Similar dependencies at different λ can be obtained experimentally. The measured data of ω and V_p can then be compared with the theoretical curves $\omega = \omega(\eta)$, $V_\mathrm{p} = V_\mathrm{p}(\eta)$ with a varying degree of accuracy which depends on the values of η. At the same time, a least square fitting procedure enables one to find the only value of η that makes this reproduction as close as possible. This value just represents the most plausible scale parameter η sought for.

It is worth recalling (see Sect. 4.4.2) that once the above four microconstants: $A_\circ$, $B_\circ$, $I_\circ$ and η, are determined no additional constants are needed for the cubic solid. The parameters $A_\circ$, $B_\circ$, $I_\circ$, η are not only necessary but also sufficient to solve any problem within the framework of DM, at any approximation of the theory. In this connection it is instructive to present here, without derivation, one of the differential equations of motion of the cubic solid in the M-th arbitrary approximation of DM ($M \geq 1$) which includes the parameters $A_\circ$, $B_\circ$, $I_\circ$ and η only:

$$\begin{aligned}
\rho\ddot{u}_1 = {} & \sum_{\kappa=1}^{M}\sum_{\zeta=1}^{M}\frac{(-1)^{\kappa-1}}{\eta^2}\Gamma_\kappa(\eta)\Gamma_\zeta(\eta)\left[A_\circ u_{1,1|\kappa+\zeta}\right. \\
& \left. + I_\circ\left(u_{1,2|\kappa+\zeta} + u_{1,3|\kappa+\zeta}\right) + B_\circ\left(u_{2,1|\kappa,2|\zeta} + u_{3,1|\kappa,3|\zeta}\right)\right] \\
& + I_\circ\sum_{\kappa=1}^{M}\frac{(-1)^{\kappa-1}}{\eta}\Gamma_\kappa(\eta)\left(\phi_{3,2|\kappa} - \phi_{2,3|\kappa}\right) \\
& + \frac{I_\circ}{2}\sum_{\kappa=1}^{M}\sum_{\zeta=1}^{M}\frac{(-1)^{\kappa-1}}{\eta}\Gamma_\kappa(\eta)\Gamma_\zeta(\eta)\left(\phi_{3,2|\kappa+\zeta} - \phi_{2,3|\kappa+\zeta}\right)
\end{aligned} \tag{4.45}$$

where

$$\Gamma_\kappa(\eta) \equiv \frac{\eta^\kappa}{\kappa!}, \qquad u_{p,q|\kappa,r|\zeta} \equiv \frac{\partial^{\kappa+\zeta}u_p}{\partial_q^\kappa \partial x_r^\zeta}, \qquad (p,q,r = 1,2,3). \tag{4.46}$$

Other symbols were given in Chap. 1. For the displacements u_2 and u_3, equations of motion are similar to eqn. (4.45). If the elastic microconstants $A_\circ$, $B_\circ$

and $I_\circ$ obey eqn. (4.40) then, in a first approximation $M = 1$, eqn. (4.45) is reduced to the Lamé equation for classical isotropic elasticity (Gould 1983):

$$\rho \ddot{u}_1 = \mu u_{1,p|2} + (\lambda + \mu) u_{p,p|1,1|1} \tag{4.47}$$

or, in conventional notation,

$$\rho \ddot{u}_1 = \mu u_{1,pp} + (\lambda + \mu) u_{p,p1}. \tag{4.48}$$

Let us note that eqn. (4.48) is obtained in view of the relations

$$\phi_2 = \frac{1}{2}(u_{1,3} - u_{3,1}), \qquad \phi_3 = \frac{1}{2}(u_{2,1} - u_{1,2}) \tag{4.49}$$

which follow from the first approximation of the couple equations (1.34) (omitted here for brevity).

5. Multi-scale, Plane Waves

V. T. Granik and M. Ferrari

5.1 Introduction

In this chapter some dynamical problems in doublet mechanics are considered which incorporate multi-scale effects. In particular, the plane propagation of elastic waves in granular media is analyzed, with reference to a spatial arrangement of the particles that results in macroscopic-level isotropy in the plane of propagation and particle displacement. The topic is motivated by experiments that establish the dependence of wave frequencies and velocities on the parameters of granular microstructure, including porosity (Urick 1948, Hampton 1967), coordination numbers and contact bonds (Trent 1989), particle sizes and interparticle distances (Iida 1938, Matsukawa and Hunter 1956). Theoretical approaches to the modeling of dispersion of plane seismological elastic waves have been proposed that incorporate some characteristic of the Earth's crust. Among these are multilayered structures (Haskell 1953), anisotropy (Crampin and Taylor 1971), gravity (Ewing et al 1957), the curvature and stratification (Sezawa 1927), radial inhomogeneity (Saito 1967), vertical discontinuities (Malischewsky 1987), etc. Other than anisotropy, these properties induce surface wave dispersion. However, the quantitative results—when provided—are in strong disagreement with observations.

In this chapter the theory of doublet mechanics is shown to predict strong dispersion, especially at shorter wavelength. However, neither this chapter nor the presented theory are dedicated to the detailed study of seismological phenomena: the emphasis is on the development of a full multiscale theory, applicable, in principle, to very different materials and microstructural dimensions. Consistent with this objective, doublet mechanics is shown below to be fully compatible with crystal dynamics and continuum elastodynamics, yet intrinsically richer than the latter in that it affords the modeling of dispersion and retardation phenomena. Both of these effects are scale-related and disappear in the continuum or the infinite wavelength limit. Doublet mechanics is compatible with lattice dynamics since these phenomena are also predicted by the latter theory.

However, doublet mechanics differs from lattice dynamics in several respects. For instance, the macroscopic elastic constants in DM are obtained through considerations of shear, torsion and elongational micromoduli as well as geometry of the underlying lattice. In lattice dynamics, one usually considers the many-body interactions in order to obtain the full set of macroscopic elastic constants as the pairwise interactions between two particles only provide some of the elastic constants (Askar 1985). Secondly, in doublet mechanics, through the introduction of microstresses one can satisfy the boundary conditions imposed on the surface of the body unlike the case in lattice dynamics where the boundary conditions, when specified, are assumed to be of periodic type (Maradudin et al 1971). Finally, the introduction of microstresses and constitutive relations at the microscopic level makes it possible to develop a general thermodynamical theory of doublet mechanics in which constitutive reductions are performed and various thermal fields are introduced which have counterparts in continuum thermoelasticity as shown in Chaps. 2 and 3. In lattice dynamics, the forces between the particles are directly specified and, as a result, such considerations do not arise at the microscopic level. More on the comparison between doublet mechanics and other theories was presented in Chap. 4.

Conventional studies of elastic wave propagation in particulate solids have neglected particle size and scaling effects. In well-developed branches of solid mechanics, such as soil dynamics (Prakash 1981), seismology (Båth 1968) and geophysics (Pilant 1979), despite the macroscopic evidence of the particulate nature of the media under study, this approach has been justified by treating only dynamic phenomena with wavelengths that are much larger than the particle dimensions.

The same reluctance to address scaling effects is also pervasive in the literature on granular media, from the earliest theories (Takahashi and Sato 1949, Takahashi and Sato 1950, Gassmann 1951) to contemporary models (Stout 1989, Wijesinghe 1989, Agarwal 1992, Ostoja-Starzewski 1992, Slade and Walton 1993).

In view of its multi-scale nature, and its capability to bridge the discrete and the continuum viewpoints, doublet mechanics offers a natural framework for the discussion of scaling effects in the dynamics of particulate and granular media as is illustrated in the subsequent sections.

5.2 Dynamic Scaling Equations

A simplified version of the doublet mechanical governing equation is now derived, with the purpose of studying mono- and bi-dimensional propagation phenomena. It is assumed that the dynamic process is isothermal and the volume forces vanish. In addition, the particle interactions are assumed to be longitudinal (central), so that the shear and torsion microstresses vanish everywhere in the body:

$$m_{\alpha i} = t_{\alpha i} \equiv 0. \tag{5.1}$$

As shown in Sects. 1.4.3 and 4.3.2, the granular medium with such properties is nonpolar: it bears only conventional (macro)stresses σ_{ij} and does not sustain couple (macro)stresses M_{ij} which are identically equal to zero in the volume V.

Moreover, the central interactions are assumed to be local, i.e., the elongation microstress p_α in an arbitrary doublet (A, B_α) depends only on its elongation microstrain ϵ_α and is independent of microstrains $\epsilon_\beta (\beta \neq \alpha)$ in the other doublets (A, B_β) originated from the same particle A. Such an interaction arises, for instance, if two particles of any doublet are supposed to be rigid and bonded by a small elastic spring. This assumption of local interaction formally means that in the constitutive relation developed in eqn. (2.78) for a nonpolar medium under isothermal conditions, the micromoduli of elasticity $A_{\alpha\beta} = A_\alpha \, \delta_{\alpha\beta}$ results in

$$p_\alpha = A_\alpha \, \epsilon_\alpha . \tag{5.2}$$

Finally, if the local interactions are homogeneous, i.e., all the micromoduli of elasticity $A_\alpha = A_\circ =$ constant for any $\alpha = 1, 2, \ldots, n$ then (5.2) yields

$$p_\alpha = A_\circ \, \epsilon_\alpha , \tag{5.3}$$

where $A_\circ$ is a microconstant of elasticity. Within such a set of assumptions, it is noted that this single microconstant suffices to fully define the physical properties of the granular medium in question.

Assumption (5.1) identically satisfyies the conservation of moment of momentum (1.56). Substituting relations (5.1) and (5.3) into the balance of linear momentum (1.55) we obtain the scale dynamic equations in terms of the displacements $u_i(x_j, t)$:

$$-A_\circ \sum_{\alpha=1}^{n} \sum_{\kappa=1}^{M} (-1)^\kappa \frac{\eta_\alpha^{\kappa-1}}{\kappa!} T_{\alpha(\kappa)} \, \overset{\circ}{\tau}_{\alpha i} \, \overset{\circ}{\tau}_{\alpha j} \sum_{\mu=1}^{M} \frac{\eta_\alpha^{\mu-1}}{\mu!} T_{\alpha(\mu)} \, \partial^{\kappa+\mu} \, u_j = \rho \ddot{u}_i , \tag{5.4}$$

where the indices $i, j = 1, 2, 3$. To simplify these equations we rewrite them as

$$- A_\circ \sum_{\alpha=1}^{n} \overset{\circ}{\tau}_{\alpha i} \, \overset{\circ}{\tau}_{\alpha j} \sum_{\kappa=1}^{M} \sum_{\mu=1}^{M} (-1)^\kappa \frac{\eta_\alpha^{\kappa+\mu-2}}{\kappa! \, \mu!} T_{\alpha(\kappa+\mu)} \, \partial^{\kappa+\mu} \, u_j = \rho \ddot{u}_i . \tag{5.5}$$

Let us now take the sum $\kappa + \mu = \delta$. Since $1 \leq \kappa \leq M$ and $1 \leq \mu \leq M$, we have $2 \leq \delta \leq 2M$. For example, if $M = 1$, then $\delta = 2$; if $M = 2$, then $\delta = 2, 3, 4$; and so on. Denoting $R \equiv 2M$, we fulfill identical transformations of the internal double sum in eqns. (5.5):

$$\begin{aligned} &\sum_{\kappa=1}^{M} \sum_{\mu=1}^{M} (-1)^\kappa \frac{\eta_\alpha^{\kappa+\mu-2}}{\kappa! \, \mu!} T_{\alpha(\kappa+\mu)} \, \partial^{\kappa+\mu} \, u_j \\ &\quad \equiv \sum_{\delta=2}^{R} \sum_{\kappa=1}^{\delta-1} (-1)^\kappa \frac{\eta_\alpha^{\delta-2}}{\kappa! \, (\delta-\kappa)!} T_{\alpha(\delta)} \, \partial^\delta \, u_j \end{aligned}$$

$$\equiv \sum_{\delta=2}^{R} \eta_\alpha^{\delta-2} T_{\alpha(\delta)} \partial^\delta u_j \sum_{\kappa=1}^{\delta-1} \frac{(-1)^\kappa}{\kappa!\,(\delta-\kappa)!} \tag{5.6}$$

The last sum in eqns. (5.6) may be computed in terms of Newton's binomial formula

$$(a+b)^\delta = \sum_{\kappa=0}^{\delta} \frac{\delta!}{\kappa!\,(\delta-\kappa)!}\, a^\kappa\, b^{\delta-\kappa} \tag{5.7}$$

which at $a=-1$, $b=1$ gives

$$\sum_{\kappa=0}^{\delta} (-1)^\kappa \frac{\delta!}{\kappa!\,(\delta-\kappa)!} \equiv 0. \tag{5.8}$$

Identity (5.8) may be rewritten as follows:

$$\sum_{\kappa=0}^{\delta} (-1)^\kappa \frac{\delta!}{\kappa!\,(\delta-\kappa)!} \equiv \delta! \sum_{\kappa=0}^{\delta} \frac{(-1)^\kappa}{\kappa!\,(\delta-\kappa)!} \tag{5.9}$$

$$\equiv \delta! \sum_{\kappa=1}^{\delta-1} \frac{(-1)^\kappa}{\kappa!\,(\delta-\kappa)!} + S \tag{5.10}$$

$$\equiv 0, \tag{5.11}$$

where $S=2$ if δ is even, and $S=0$ if δ is odd. The relations (5.10) and (5.11) yields

$$\sum_{\kappa=1}^{\delta-1} \frac{(-1)^\kappa}{\kappa!\,(\delta-\kappa)!} \equiv -(\frac{S}{\delta!}). \tag{5.12}$$

On substituting the identity (5.12) into eqns. (5.6) and then into eqns. (5.5), we finally obtain the following three basic equations of scaling microdynamics:

$$\rho \ddot{u}_i = 2\,A_\circ \sum_{\alpha=1}^{n} \overset{\circ}{\tau}_{\alpha i}\, \overset{\circ}{\tau}_{\alpha j} \sum_{\delta=2,4,\ldots}^{R} \frac{\eta_\alpha^{\delta-2}}{\delta!}\, T_{\alpha(\delta)}\, \partial^\delta\, u_j. \tag{5.13}$$

These equations include the scaling parameters η_α in an explicit form. In the first (or nonscale) approximation, i.e., for $M=1$, the scaling parameters η_α vanish, and eqns. (5.13) reduce to

$$\rho \frac{\partial^2 u_i}{\partial t^2} = A_\circ \sum_{\alpha=1}^{n} \overset{\circ}{\tau}_{\alpha i}\, \overset{\circ}{\tau}_{\alpha j}\, \overset{\circ}{\tau}_{\alpha k}\, \overset{\circ}{\tau}_{\alpha l} \frac{\partial^2 u_j}{\partial x_k\, \partial x_l}. \tag{5.14}$$

For clarity, the multi-scale equations (5.13) may be written *in extenso* as

$$\rho \frac{\partial^2 u_i}{\partial t^2} = 2A_\circ \sum_{\alpha=1}^{n} \overset{\circ}{\tau}_{\alpha i} \overset{\circ}{\tau}_{\alpha j} \Bigg(\frac{1}{2!}\, \overset{\circ}{\tau}_{\alpha k}\, \overset{\circ}{\tau}_{\alpha l} \frac{\partial^2 u_j}{\partial x_k \partial x_l}$$

$$+ \frac{\eta_\alpha^2}{4!}\, \overset{\circ}{\tau}_{\alpha k}\, \overset{\circ}{\tau}_{\alpha l}\, \overset{\circ}{\tau}_{\alpha p}\, \overset{\circ}{\tau}_{\alpha q} \frac{\partial^4 u_j}{\partial x_k\, \partial x_l\, \partial x_p\, \partial x_q}$$

$$+ \frac{\eta_\alpha^4}{6!} \overset{\circ}{\tau}_{\alpha k} \overset{\circ}{\tau}_{\alpha l} \overset{\circ}{\tau}_{\alpha p} \overset{\circ}{\tau}_{\alpha q} \overset{\circ}{\tau}_{\alpha r} \overset{\circ}{\tau}_{\alpha s} \frac{\partial^6 u_j}{\partial x_k \, \partial x_l \, \partial x_p \, \partial x_q \, \partial x_r \, \partial x_s}$$

$$+ \cdots \Big), \tag{5.15}$$

or, in an explicit tensor notation,

$$\rho \frac{\partial^2 u_i}{\partial t^2} = \sum_{\kappa=2,4,\ldots}^{R=2M} C_{ijk_1 \ldots k_\kappa} \frac{\partial^\kappa u_j}{\partial x_{k_1} \cdots \partial x_{k_\kappa}}, \tag{5.16}$$

where

$$C_{ijk_1 \ldots k_\kappa} \equiv 2 A_\circ \sum_{\alpha=1}^{n} \frac{\eta_\alpha^{\kappa-2}}{\kappa!} \overset{\circ}{\tau}_{\alpha i} \overset{\circ}{\tau}_{\alpha j} \overset{\circ}{\tau}_{\alpha k_1} \cdots \overset{\circ}{\tau}_{\alpha k_\kappa} \tag{5.17}$$

are tensors of rank $\kappa + 2$ that correspond to generalized moduli of elasticity. Formula (5.17) shows that these moduli comprise micro-level constitutive information ($A_\circ$), as well as microstructural parameters ($\eta_\alpha, \overset{\circ}{\tau}_{\alpha i}$). In the first approximation, the scaling dynamic equations (5.16) are form-identical with the non-scaling equations of the conventional theory of anisotropic elasticity (Pilant 1979)

$$\rho \frac{\partial^2 u_i}{\partial t^2} = C_{ijkl} \frac{\partial^2 u_j}{\partial x_k \, \partial x_l}. \tag{5.18}$$

Within the doublet mechanical context, however, the stiffness tensor is

$$C_{ijkl} \equiv A_\circ \sum_{\alpha=1}^{n} \overset{\circ}{\tau}_{\alpha i} \overset{\circ}{\tau}_{\alpha j} \overset{\circ}{\tau}_{\alpha k} \overset{\circ}{\tau}_{\alpha l}. \tag{5.19}$$

Equations (5.18) and (5.19) are equivalent to eqns. (5.14). There is a major difference between the conventional understanding of eqn. (5.18) and its doublet mechanical counterpart; the tensor C_{ijkl} in eqn. (5.19) is invariant with respect to any permutation of its subscripts, including $C_{ijkl} = C_{ikjl}$ and thus possesses only 15 independent constants: C_{1231}, C_{1232}, C_{1233}, C_{1211}, C_{1212}, C_{1222}, C_{1311}, C_{1313}, C_{1333}, C_{2322}, C_{2323}, C_{2333}, C_{1111}, C_{2222}, and C_{3333}. By contrast, the number of independent constants in the conventional theory of anisotropic elasticity is 21.

In order for an arbitrary anisotropic elastic granular medium to be completely characterized by 15 independent constants it is sufficient that the microforces of interaction between constitutive particles (granules, molecules, etc.) be central, local, and homogeneous in the sense specified above. It can easily be shown that the first two conditions are also necessary. The third condition is not necessary and was adopted here because it has eventually led to the simplest form of scaling equations (5.15)–(5.17). In the case of isotropy, the elastic features are characterized by one macroconstant of elasticity (see below eqn. (5.68)). We further remark that a and similar result is obtained using the approach of lattice dynamics where the interparticle forces are assumed to be central (Askar 1985). However, in DM, using the above governing equations, it can also be shown that the tensor C_{ijkl} has 21 independent components if the interparticle microforces are only central. The basic equations of scaling microdynamics become then much more complicated than eqns. (5.15)–(5.17). The proofs of the above statements are omitted for brevity. The reader interested in the historical discussion of the multiconstancy versus rariconstancy is referred to (Todhunter 1886, articles 921–934).

Equation (5.17) may be interpreted in two, quite different manners. In the first, the microstructural parameters $A_\circ$, η_α and $\tau^\circ_{\alpha i}$ are assumed known, and formula (5.17) enables one to compute the microstructural tensors $C_{ijk_1...k_\kappa}$ and then apply the dynamic equations in explicit tensor notation, eqns. (5.16). However, computing the tensors $C_{ijk_1...k_\kappa}$ is not a necessary step and may be avoided by inserting the known microstructural parameters directly into the dynamic equations (5.15). Alternatively, if the microstructure of a granular body is unknown, the elastic moduli $C_{ijk_1...k_\kappa}$ must be considered as some macroscopic parameters of the body to be determined by special macroscopic experiments. Nevertheless, based on the underlying theory, the experimental variables are only $A_\circ$, η_α, $\tau^\circ_{\alpha i}$.

Thus, using the doublet microstructural approach, we have first obtained the dynamic scaling equations (5.15)–(5.17). These equations not only enable us to study dynamic scaling problems for granular and particulate materials but also allow us to do this, if necessary, from two quite different viewpoints: microscopic (eqn. (5.15)) and macroscopic (eqns. (5.16), (5.17)).

5.3 Plane Elastic Waves in Granular Media

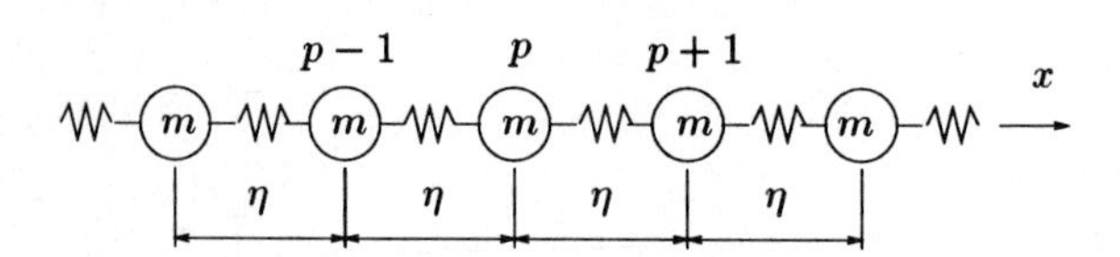

Fig. 5.1. A linear monatomic lattice

We begin by studying the propagation of waves in a one dimensional lattice and subsequently consider plane waves in granular media. To this end, consider a linear monatomic lattice with the valence $n = 1$ and internodal distance $\eta_\alpha \equiv \eta =$ constant (Fig. 5.1). The only lattice direction which corresponds to $\alpha = 1$ is taken to be parallel to the Cartesian axis $x_1 \equiv x$ so that the lattice axis $\boldsymbol{\tau}^\circ_\alpha = \mathbf{e}_1$. Since $\boldsymbol{\tau}^\circ_\alpha = \tau^\circ_{\alpha i}\,\mathbf{e}_i$, it follows that

$$\tau^\circ_{11} = 1 \text{ and } \tau^\circ_{\alpha i} = 0 \text{ if } \alpha \neq 1 \text{ or } i \neq 1. \tag{5.20}$$

The relations (5.17) and (5.20) then imply

$$C_{ijk_1\dots k_\kappa} = \begin{cases} 2\,A_\circ\,\eta^{\kappa-2}/\kappa! & \text{if } i=j=k_1=\dots=k_\kappa=1 \\ 0 & \text{otherwise.} \end{cases} \tag{5.21}$$

In view of eqn. (5.21), the dynamic equation (5.16) becomes

$$\rho\,\frac{\partial^2 u}{\partial t^2} = 2\,A_\circ \sum_{\kappa=2,4,\dots}^{R=2M} \frac{\eta^{\kappa-2}}{\kappa!}\,\frac{\partial^\kappa u}{\partial x^\kappa} = \frac{2\,A_\circ}{\eta^2} \sum_{\kappa=2,4,\dots}^{R=2M} \frac{\eta^\kappa}{\kappa!}\,\frac{\partial^\kappa u}{\partial x^\kappa}, \tag{5.22}$$

in which $u = u_1$ is a node displacement along the x-axis.

It is of interest now to demonstrate that the differential equation (5.22) actually contains the difference equation employed in solid state physics for wave propagation in a monoatomic lattice. To this end, the limit of the right-hand side of eqn. (5.22) is taken for $M \to \infty$, yielding

$$\sum_{\kappa=2,4,\dots}^{\infty} \frac{\eta_\alpha^\kappa}{\kappa!}\,\frac{\partial^\kappa u}{\partial x^\kappa} \equiv \frac{1}{2}(u_{p+1} + u_{p-1} - 2u_p) \tag{5.23}$$

where $u_p \equiv u \equiv u(x,t)$, $u_{p-1} \equiv u(x-\eta,t)$, $u_{p+1} \equiv u(x+\eta,t)$ denote the displacements of an arbitrary p-th node and its nearest neighbors to the left and the right, respectively (see Fig. 5.1). To prove eqn. (5.23) let the displacements u_{p-1} and u_{p+1} be expanded into Taylor's series about x:

$$u_{p-1} \equiv u(x-\eta,t) \equiv u + \sum_{\kappa=1}^{\infty} (-1)^\kappa \frac{\eta_\alpha^\kappa}{\kappa!}\,\frac{\partial^\kappa u}{\partial x^\kappa}, \tag{5.24}$$

$$u_{p+1} \equiv u(x+\eta,t) \equiv u + \sum_{\kappa=1}^{\infty} \frac{\eta_\alpha^\kappa}{\kappa!}\,\frac{\partial^\kappa u}{\partial x^\kappa}. \tag{5.25}$$

The sum of series (5.24) and (5.25) is

$$u_{p+1} + u_{p-1} \equiv 2\left(u + \sum_{\kappa=2,4,\dots}^{\infty} \frac{\eta_\alpha^\kappa}{\kappa!}\,\frac{\partial^\kappa u}{\partial x^\kappa}\right). \tag{5.26}$$

Since $u \equiv u_p$, the relation (5.26) directly entails the identity (5.23), as was to be shown. Substituting eqn. (5.26) into the differential equation (5.22) at $M \to \infty$, we obtain the difference equation of motion of an arbitrary p-th node, or atom:

$$m\,\frac{\partial^2 u_p}{\partial t^2} = C\,(u_{p+1} + u_{p-1} - 2u_p), \tag{5.27}$$

where $m = \rho\eta^3$ may be interpreted as the mass of an atom in the monatomic lattice, and $C = A_\circ\eta$ is the force constant. Equation (5.27) may be found in textbooks on solid state physics (Kittel 1986) and in the dynamics of atoms in crystals (Cochran 1973).

Consider now a longitudinal wave

$$u_p = u_\circ \exp[i\,(\omega t - kp\eta)] \tag{5.28}$$

traveling along the axis x to the right. Here, $i = \sqrt{-1}$, while $u_\circ$, ω, k are the wave amplitude, (angular) frequency, and wavenumber, respectively. The quantity $p\eta$ is a discrete analog of the continuous variable x. On substituting the function u_p from eqn. (5.28) into the dynamic equation (5.27) we find u_p to be a solution of eqn. (5.27) provided that

$$-m\omega^2 = C\,[\exp(ik\eta) + \exp(-ik\eta) - 2]. \tag{5.29}$$

Since $\exp(ik\eta) \equiv \cos(k\eta) + i\,\sin(k\eta)$, eqn. (5.29) becomes

$$\omega = 2\sqrt{\frac{C}{m}}\,\sin\left(\frac{k\eta}{2}\right) \equiv \frac{2}{\eta}\sqrt{\frac{A_\circ}{\rho}}\,\sin\left(\frac{k\eta}{2}\right). \tag{5.30}$$

In view of the identity $k = 2\pi/\lambda$, λ being the wavelength, eqn. (5.30) yields the expression for the phase velocity V_p:

$$V_\mathrm{p} \equiv \frac{\omega}{k} = V_\circ\,\frac{\sin(\pi\eta/\lambda)}{\pi\eta/\lambda}, \tag{5.31}$$

where $V_\circ \equiv \sqrt{A_\circ/\rho}$ is the phase velocity for waves of infinite wavelength. The case of infinitely long waves may be alternatively obtained by restricting the general dynamic equation (5.22) to the first, nonscale, approximation ($M = 1$, $R = 2$)

$$\rho\,\frac{\partial^2 u}{\partial t^2} = A_\circ\,\frac{\partial^2 u}{\partial x^2} \tag{5.32}$$

and considering the continuous analog to the longitudinal wave (5.28)

$$u = u_\circ\,\exp[i(\omega t - kx)]. \tag{5.33}$$

Inserting eqn. (5.33) into eqn. (5.32), we obtain the frequency $\omega = k\sqrt{A_\circ/\rho}$ and the phase velocity $V_\mathrm{p} \equiv \omega/k = \sqrt{A_\circ/\rho} \equiv V_\circ$. The quantity $V_\circ$ is known to be the velocity of sound in a classical—nonscale—continuum (Cochran 1973).

The function $\omega(k)$ in eqn. (5.30) is periodic with periodicity $k = 2\pi/\eta$. Therefore with no loss of generality the wavenumber k may be restricted to the range $\pm\pi/\eta$: if $0 < k \leq \pi/\eta$, the wave travels to the right, and if $-\pi/\eta \leq k < 0$, the wave travels to the left. Since $|k| \leq \pi/\eta$ and $\lambda \equiv 2\pi/k$, it follows that

$$\lambda \geq 2\eta, \tag{5.34}$$

i.e., in the monatomic structure, the only wave modes that may propagate are those of the wavelength λ no smaller than twice the internodal distance (or twice the doublet length) 2η. The restriction (5.34) holds for all particulate materials.

According to eqn. (5.30), the phase velocity V_p depends on the wavelength λ. This means that longitudinal waves in the monatomic lattice are dispersive in contrast to those modeled by the conventional nonscale wave equation (5.32) which are known to be nondispersive. By eqn. (5.31), the velocity V_p

reaches the maximum $V_{\mathrm{p,max}} = V_\circ \equiv \sqrt{A_\circ/\rho}$ at the maximum wavelength $\lambda_{\max} \to \infty$, and the minimum $V_{\mathrm{p,min}} = 2\,V_\circ/\pi \doteq 0.63662\,V_\circ$ at the minimum wavelength $\lambda_{\min} = 2\eta$. This scale effect may be called the wave retardation at short wavelengths.

Next, consider a plane wave

$$u_j = u_{j\circ}\,\exp[i(\omega t - kx)], \tag{5.35}$$

traveling in an unbounded granular medium along the axis $x \equiv x_1$. The symbols i, ω, k are the same as in eqn. (5.33). It follows from the expression (5.35) that

$$\rho\,\frac{\partial^2 u_j}{\partial t^2} = -\omega^2\,u_{j\circ}\,\exp[i(\omega t - kx)], \tag{5.36}$$

$$\frac{\partial^\kappa\,u_j}{\partial x_1^\kappa} \equiv \frac{\partial^\kappa u_j}{\partial x^\kappa} = (-1)^{\kappa/2}\,k^\kappa\,u_{j\circ}\,\exp[i(\omega t - kx)], \tag{5.37}$$

$$\frac{\partial^\kappa u_j}{\partial x_2^\kappa} = \frac{\partial^\kappa u_j}{\partial x_3^\kappa} \equiv 0, \tag{5.38}$$

where $\kappa = 2, 4, 6, \ldots$. Substituting eqns. (5.36)–(5.38) into eqn. (5.16) and using eqn. (5.17), we obtain the following set of homogeneous algebraic equations:

$$u_{j\circ}\,\omega^2 + u_{l\circ}\,V_\circ^2\,F_{jl} = 0, \tag{5.39}$$

in which

$$F_{jl} \equiv F_{lj} = 2 \sum_{\kappa=2,4,\ldots}^{R=2M} (-1)^{\kappa/2}\,\frac{k^\kappa}{\kappa!} \sum_{\alpha=1}^{n} \eta_\alpha^{\kappa-2}\,\tau_{\alpha j}^\circ\,\tau_{\alpha l}^\circ\,(\tau_{\alpha 1}^\circ)^\kappa\,. \tag{5.40}$$

We assume that the $x_3 \equiv z$-component of the displacement $\mathbf{u} = u_j\,\mathbf{e}_j$ vanishes in all the medium in question, so that $u_3 \equiv 0$. The medium is therefore in a state of plane-strain and the indices $j, l = 1, 2$ everywhere in this section. The same simplification would obviously result from considering a three-dimensional medium, with translational symmetry of the hexagonal type. Under either assumption, the above set (5.39) reduces to a system of two homogeneous equations, which admits a nonzero solution if and only if its determinant is equal to zero (Sokolnikoff and Sokolnikoff 1941):

$$\begin{vmatrix} \omega^2 + V_\circ^2\,F_{11} & V_\circ^2\,F_{12} \\ V_\circ^2\,F_{21} & \omega^2 + V_\circ^2\,F_{22} \end{vmatrix} = 0 \tag{5.41}$$

where $V_\circ \equiv \sqrt{A_\circ/\rho}$. The parameters F_{jl} are functions of the lattice directions $\tau_{\alpha j}^\circ$ and the doublet length η_α and thus depend on the microstructure of the granular medium. Therefore the solution of eqn. (5.41) also depends on the microstructure. We are going to get a numerical solution of eqn. (5.41) which demands focusing attention on a particular microstructure. So we will further

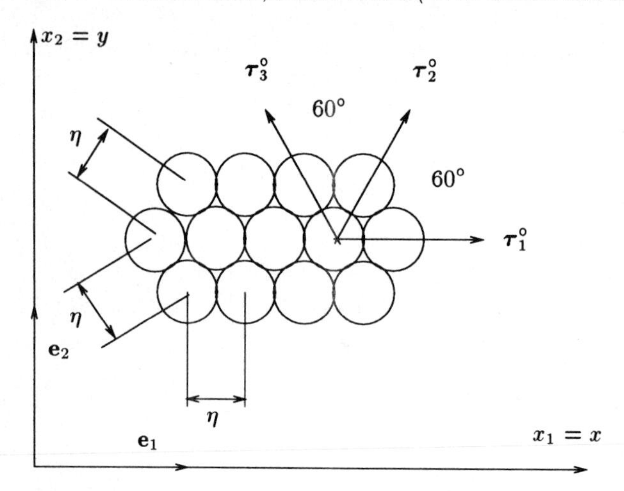

Fig. 5.2. A cubical-tetrahedral packing H_4

consider the cubical-tetrahedral packing H_4 which in the plane $(x_1, x_2) \equiv (x, y)$ resembles a honeycomb pattern (Fig. 5.2).

The packing H_4 has in general the valence $n = 4$, i.e., four spatial directions that are determined by the following four unit vectors (see Fig. 5.2):

$$\left.\begin{array}{ll} \boldsymbol{\tau}_1^\circ = \mathbf{e}_1, & \boldsymbol{\tau}_2^\circ = \mathbf{e}_1 \cos\varphi + \mathbf{e}_2 \sin\varphi, \\ \boldsymbol{\tau}_3^\circ = -\mathbf{e}_1 \cos\varphi + \mathbf{e}_2 \sin\varphi, & \boldsymbol{\tau}_4^\circ = \mathbf{e}_3, \end{array}\right\} \tag{5.42}$$

with the structural angle $\varphi = 60°$. Since the translation $u_3 \equiv 0$ and we only consider the wave displacements u_1 and u_2 in the plane (x_1, x_2), the fourth microstructure direction $\boldsymbol{\tau}_4^\circ = \mathbf{e}_3$ which is parallel to the $x_3 \equiv z$-axis and perpendicular to the plane (x_1, x_2) does not play any role in this problem. By eqn. (5.42), the direction cosine matrix is

$$[\tau_{\alpha j}^\circ] = \begin{bmatrix} \tau_{11}^\circ & \tau_{12}^\circ \\ \tau_{21}^\circ & \tau_{22}^\circ \\ \tau_{31}^\circ & \tau_{32}^\circ \end{bmatrix} = \begin{bmatrix} 1 & 0 \\ 1/2 & \sqrt{3/2} \\ -1/2 & \sqrt{3/2} \end{bmatrix} \tag{5.43}$$

Equation (5.40) takes the form

$$F_{jl} \equiv F_{lj} = \sum_{\kappa=2,4,\ldots}^{R=2M} (-1)^{\kappa/2} \frac{2}{\eta^2} \frac{(k\eta)^\kappa}{\kappa!} a_{jl}, \tag{5.44}$$

where $\eta_\alpha = \eta =$ constant for all $\alpha = 1, 2, 3$ and

$$a_{jl} \equiv a_{lj} = \sum_{\alpha=1}^{n} \tau_{\alpha j}^\circ \tau_{\alpha l}^\circ (\tau_{\alpha 1}^\circ)^\kappa . \tag{5.45}$$

According to eqns. (5.43) and (5.45), the parameters a_{jl} are determined by the formulas

$$a_{11} = 1 + 2^{-(\kappa+1)}, \qquad a_{22} = 3 \cdot 2^{-(\kappa+1)}, \qquad a_{12} = a_{21} = 0. \tag{5.46}$$

Substituting these expressions of a_{jl} into eqn. (5.44), we obtain

$$F_{11} = 2\,k^2 \sum_{\kappa=2,4,\ldots}^{R=2M} (-1)^{\kappa/2} \frac{(k\eta)^{\kappa-2}}{\kappa!} [1 + 2^{-(\kappa+1)}], \tag{5.47}$$

$$F_{22} = 6\,k^2 \sum_{\kappa=2,4,\ldots}^{R=2M} (-1)^{\kappa/2} \frac{(k\eta)^{\kappa-2}}{\kappa!}\, 2^{-(\kappa+1)}, \tag{5.48}$$

The parameters $F_{12} \equiv F_{21} = 0$ and thus eqn. (5.41) splits up into two separate dispersion equations:

$$\omega^2 + V_\circ^2\, F_{11} = 0, \tag{5.49}$$

$$\omega^2 + V_\circ^2\, F_{22} = 0. \tag{5.50}$$

Equation (5.49) involves the parameter F_{11} and by eqns. (5.39) and (5.35) concerns the displacement u_1 that is parallel to the x_1-axis; eqn. (5.50) includes the parameter F_{22} and according to eqn. (5.39) and eqn. (5.35) concerns the displacement u_2 that is perpendicular to the x_1-axis. Since we consider the plane wave (5.35) traveling along the axis x_1, eqn. (5.49) relates to longitudinal elastic displacements, or P-waves, and eqn. (5.50) refers to transverse elastic displacements, or S-waves. The fact that the P-waves and the S-waves are described by separate equations means that these waves are not interconnected and propagate quite independently of each other. The characteristics of P-waves (amplitudes, frequencies, velocities) are independent of the same characteristics of S-waves—exactly as in the classical elastic isotropic continuum (Kolsky 1963).

The phase velocities V_{pp} and V_{ps} of the P- and S-waves, respectively, are found via eqns. (5.47)–(5.50) to be:

$$V_{\text{pp}} \equiv \frac{\omega}{k} = \frac{1}{2} V_\circ \left[\sum_{\kappa=2,4,\ldots}^{R=2M} (-1)^{1+\kappa/2} \left(1 + 2^{\kappa+1}\right) \frac{\pi^{\kappa-2}}{\kappa!} \left(\frac{\eta}{\lambda}\right)^{\kappa-2} \right]^{1/2}, \tag{5.51}$$

$$V_{\text{ps}} \equiv \frac{\omega}{k} = \frac{\sqrt{3}}{2} V_\circ \left[\sum_{\kappa=2,4,\ldots}^{R=2M} (-1)^{1+\kappa/2} \frac{\pi^{\kappa-2}}{\kappa!} \left(\frac{\eta}{\lambda}\right)^{\kappa-2} \right]^{1/2} \tag{5.52}$$

where the identity $k = 2\pi/\lambda$ has been taken into account, λ being the wavelength.

The group velocities V_{g} are related to the phase velocities V_{p} as (Brillouin 1960)

$$V_{\text{g}} \equiv \frac{\partial \omega}{\partial k} = V_{\text{p}} - \lambda \frac{\partial V_{\text{p}}}{\partial \lambda}. \tag{5.53}$$

By substituting the phase velocities V_{pp} and V_{ps} from eqns. (5.51)–(5.52) into eqn. (5.53) we obtain the group velocities V_{gp} and V_{gs} of the P- and S-waves, respectively:

$$V_{gp} = V_{pp} - \frac{1}{8\,V_{pp}} \sum_{\kappa=2,4,\ldots}^{R=2M} (-1)^{\kappa/2}\,(1+2^{\kappa+1})\,\frac{(\kappa-2)\pi^{\kappa-2}}{\kappa!}\left(\frac{\eta}{\lambda}\right)^{\kappa-2}, \tag{5.54}$$

$$V_{gs} = V_{ps} - \frac{3}{8\,V_{ps}} \sum_{\kappa=2,4,\ldots}^{R=2M} (-1)^{\kappa/2}\,\frac{(\kappa-2)\pi^{\kappa-2}}{\kappa!}\left(\frac{\eta}{\lambda}\right)^{\kappa-2}. \tag{5.55}$$

The formulas (5.51)–(5.55) express the wave velocities to an arbitrary M-th approximation for M from 1 to ∞. In the first approximation ($M = 1$), the above formulas yield the following expressions for the wave velocities:

$$V_{pp}^{(1)} = V_{gp}^{(1)} = V_{op} \equiv \frac{3}{2\sqrt{2}}\,V_{\circ}, \tag{5.56}$$

$$V_{ps}^{(1)} = V_{gs}^{(1)} = V_{os} \equiv \frac{\sqrt{3}}{2\sqrt{2}}\,V_{\circ}, \tag{5.57}$$

which are independent of the nondimensional scale parameter η/λ. Alternatively, these expressions may be arrived at via two different assumptions:

(A) the constituent granules of the solid are material points (whose sizes are infinitesimal: $\eta = 0$) and all the waves may have arbitrary but finite length λ, or

(B) the constituent particles of the solid may have an arbitrary but finite size $\eta \neq 0$ and all the waves have an infinite length $\lambda \to \infty$.

The assumption (A) is adopted in classical theory of elasticity, while the restriction (B) is employed in the special discrete-continuum theories of wave propagation in granular media (Takahashi and Sato 1949, 1950, Gassmann 1951) and crystals (Cochran 1973). It is clear that neither of these approaches permits the analysis of scale effects.

The question of convergence of the series appearing in the wave velocity expressions (5.51)–(5.55) is considered next. Consider eqn. (5.51) first, in the limit as $M \to \infty$. Squaring both sides of the relation, we obtain an infinite alternating series on the right-hand side of eqn. (5.51). According to Leibnitz's theorem (Sokolnikoff and Sokolnikoff 1941), such series is convergent if its terms satisfy the following conditions:

$$a_{\kappa+2} < a_{\kappa}, \tag{5.58}$$

$$\lim_{\kappa\to\infty} a_{\kappa} = 0. \tag{5.59}$$

In the inequality (5.58), we took into account that the term $a_{\kappa+2}$ follows a_{κ} because a subscript κ only runs through even values $2, 4, \ldots$. The relation (5.51) shows that

$$a_\kappa = \left(1 + 2^{\kappa+1}\right) \frac{\pi^{\kappa-2}}{\kappa!} \left(\frac{\eta}{\lambda}\right)^{\kappa-2}, \tag{5.60}$$

$$a_{\kappa+2} = \left(1 + 2^{\kappa+3}\right) \frac{\pi^{\kappa}}{(\kappa+2)!} \left(\frac{\eta}{\lambda}\right)^{\kappa}. \tag{5.61}$$

Substituting a_κ and $a_{\kappa+2}$ from eqns. (5.60) and (5.61) into eqn. (5.58), we transform it to the inequality

$$\pi^2 \left(\frac{\eta}{\lambda}\right)^2 < \frac{1 + 2^{\kappa+1}}{1 + 2^{\kappa+3}} (\kappa+1)(\kappa+2). \tag{5.62}$$

According to eqn. (5.34), $\max(\eta/\lambda) = 1/2$, and therefore the left-hand side of eqn. (5.62) has $\max\left[\pi^2 (\eta/\lambda)^2\right] = \pi^2/4 \doteq 2.467401$. On the other hand, at $\kappa = 2$, the right-hand side of eqn. (5.62) has $\min\left[\frac{1+2^{\kappa+1}}{1+2^{\kappa+3}} (\kappa+1)(\kappa+2)\right] = 3.272727 > 2.467401$. Thus inequality (5.62) and, consequently, the condition (5.58) are satisfied.

We now turn to eqn. (5.61) and use Stirling's formula (Sokolnikoff and Sokolnikoff 1941)

$$\kappa! \simeq \sqrt{(2\pi\kappa)} \left(\frac{\kappa}{e}\right)^{\kappa}. \tag{5.63}$$

Substituting eqn. (5.63) into eqn. (5.59) and taking account of the obvious inequalities $1 + 2^{\kappa+1} < 2^{\kappa+2}$, $\sqrt{2\pi\kappa} > 1$, $\eta/\lambda < 1$, we obtain

$$a_\kappa = \frac{\left(1 + 2^{\kappa+1}\right) e^{\kappa}}{\sqrt{(2\pi\kappa)}\, \kappa^{\kappa}} \frac{\pi^{\kappa}}{\pi^2} \left(\frac{\eta}{\lambda}\right)^{\kappa-2} < \frac{4}{\pi^2} \left(\frac{2\pi e}{\kappa}\right)^{\kappa} < \left(\frac{2\pi e}{\kappa}\right)^{\kappa}. \tag{5.64}$$

Since $\lim_{\kappa\to\infty} (2\pi e/\kappa)^{\kappa} = 0$, the term a_κ, by eqn. (5.64), obeys the condition (5.59). Thus the infinite series in eqn. (5.51) is convergent.

The convergence of the other three series in relations the (5.52), (5.54) and (5.55), with $M \to \infty$, is proven analogously. Because of convergence, the sums of these infinite alternating series may be determined by truncating them and computing the remainder, which is always less than the first truncated term a_{R+2} (Sokolnikoff and Sokolnikoff 1941).

Let us, for instance, calculate the remainder Δ for the infinite series in eqn. (5.51) by using the formula (5.61) and replacing κ by R:

$$\Delta \le a_{R+2} \equiv \left(1 + 2^{R+3}\right) \frac{\pi^{R}}{(R+2)!} \left(\frac{\eta}{\lambda}\right)^{R}. \tag{5.65}$$

The relation (5.65) shows that the term a_{R+2} attains a maximum when η/λ reaches a maximum which, by eqn. (5.34), is equal to 1/2: then $\max a_{R+2} = 2.497 \times 10^{-20}$ ($R = 30$, $M = 15$). In this case, the remainder Δ also attains a maximum that is no larger than 2.497×10^{-20}, and hence $\max\sqrt{\Delta} \le 1.58010^{-10} < 10^{-9}$. Thus, if we truncate the infinite series in eqn. (5.52) at the 15-th term ($M = 15$, $R = 30$), we approximate the phase velocity V_{pp} to

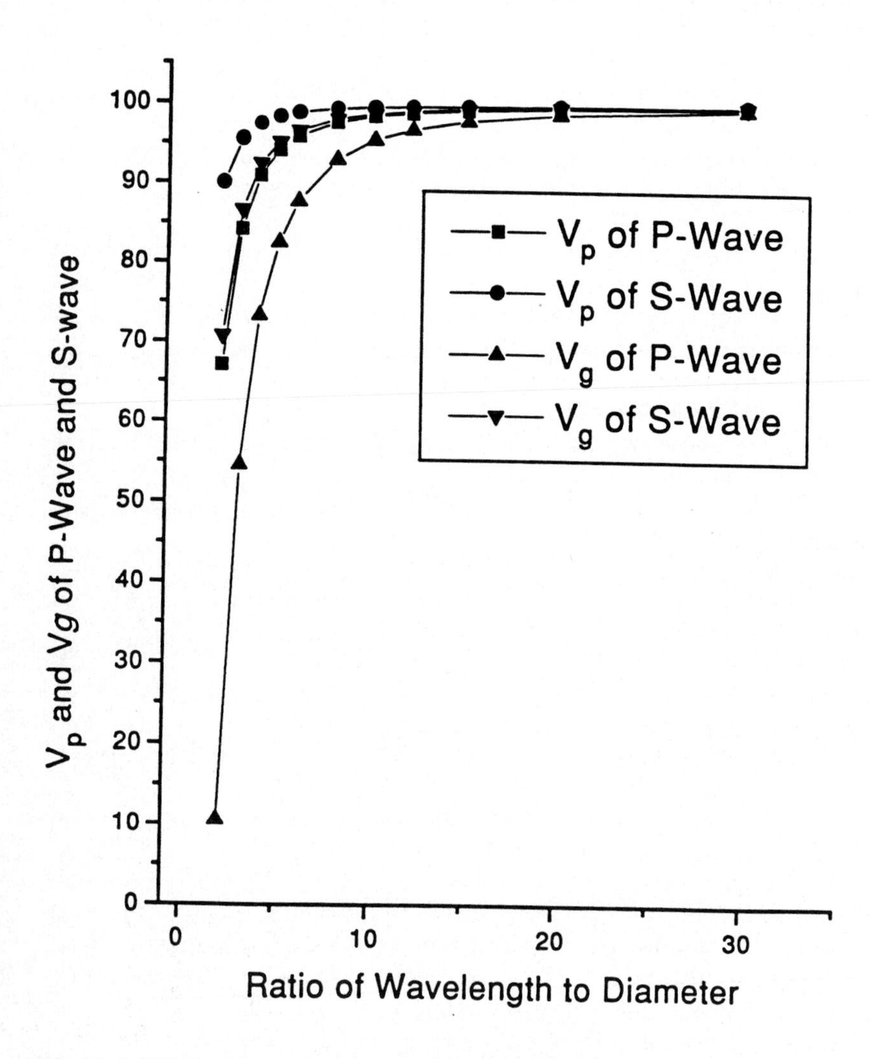

Fig. 5.3. Wave velocities versus normalized wavelength

within one part in a billion. A completely analogous analysis can be carried out for the other series given in equations (5.52), (5.54) and (5.55).

The results of the calculations are presented in Fig. 5.3, which compares the phase and group velocities of P- and S-waves in scale elastic granular media (which have finite sizes of particles $d = \eta$) with the corresponding velocities V_{op} and V_{os} in conventional nonscale elastic media (where $\eta = 0$) that are determined by classical theory of elasticity (see, also, eqns. (5.56)–(5.57)).

Figure 5.3 shows a most interesting feature of the plane P- and S-waves in scale granular media. The waves have dispersion because all their velocities depend on the wavelength λ and the granule size η. In particular, the velocities diminish as the nondimensional parameter $l = \lambda/eta$ decreases. As in the case of monoatomic crystals we observe the phenomenon of wave retardation at relatively short wavelengths.

It is further noted that, in the bands of relatively long waves when the parameter $l \equiv \lambda/\eta$ is large enough, i.e., $l > 15$ or 20, the velocities V_{pp}, V_{gp} are only slightly less than V_{op}, as well as the velocities V_{ps}, V_{gs} are somewhat less than V_{os}; the difference between these is no more than 2%.

On the other hand, the shorter the waves, the more significant the wave retardation. In the band of short wavelengths that are close to their minimum—a double particle diameter (when $l = 2$)—the velocities reach the least values that are considerably less than V_{op} and V_{os}. In this band, classical theory of elasticity significantly overestimates the wave velocities: phase velocities up to 33% for P-waves and 10% for S-waves, and group velocities up to 89% for P-waves and 29% for S-waves.

5.4 Discussion and Closure

In the previous section, the doublet elastodynamics of the basal plane of the cubical-tetrahedral packing H_4 was considered, with the purpose of establishing the fact that elastic plane waves in an isotropic granular medium are dispersive. It was thereby demonstrated that the capability of modeling dispersion is lost upon introducing the long-wavelength or the continuum approximations. In analogy with isotropic continuum elastodynamics, it was also shown that the longitudinal and shear waves are decoupled. To place this analogy in the proper perspective, however, it is remarked that the basal plane of H_4 is isotropic only at the nonscale approximation ($M = 1$). As a side remark, it is noted that eqns. (5.18)–(5.19) do not allow for the propagation of nontrivial plane waves polarized in the normal direction (i.e., SH-waves).

Returing to the question of the scale-dependence of isotropy, we have computed the generalized macromoduli of elasticity $C_{ijk_1\ldots k_\kappa}$ in accordance with eqn. (5.17), where of $\eta_\alpha = \eta =$ constant, $\kappa = 2, 4, 6, \ldots$. For ease of representation, the associated nondimensional macromoduli $\bar{C}_{ijk_1\ldots k_\kappa}$ are introduced that are defined as

$$\bar{C}_{ijk_1\ldots k_\kappa} \equiv \frac{\kappa!\, C_{ijk_1\ldots k_\kappa}}{2A_\circ \eta^{\kappa-2}} \equiv \sum_{\alpha=1}^{n} \overset{\circ}{\tau}_{\alpha i}\, \overset{\circ}{\tau}_{\alpha j}\, \overset{\circ}{\tau}_{\alpha k_1} \cdots \overset{\circ}{\tau}_{\alpha k_\kappa}. \tag{5.66}$$

In order to establish the conclusion, the case $i = j = k_1 = \ldots = k_\kappa = 1$, is considered for different values of κ and in variuos frames differing by the angle γ, i.e., under

$$\left[\overset{\circ}{\tau}_{\alpha j}\right] = \begin{bmatrix} \overset{\circ}{\tau}_{11} & \overset{\circ}{\tau}_{12} \\ \overset{\circ}{\tau}_{21} & \overset{\circ}{\tau}_{22} \\ \overset{\circ}{\tau}_{31} & \overset{\circ}{\tau}_{32} \end{bmatrix} = \begin{bmatrix} \cos\gamma & -\sin\gamma \\ \cos(60-\gamma) & \sin(60-\gamma) \\ -\cos(60+\gamma) & \sin(60+\gamma) \end{bmatrix} \tag{5.67}$$

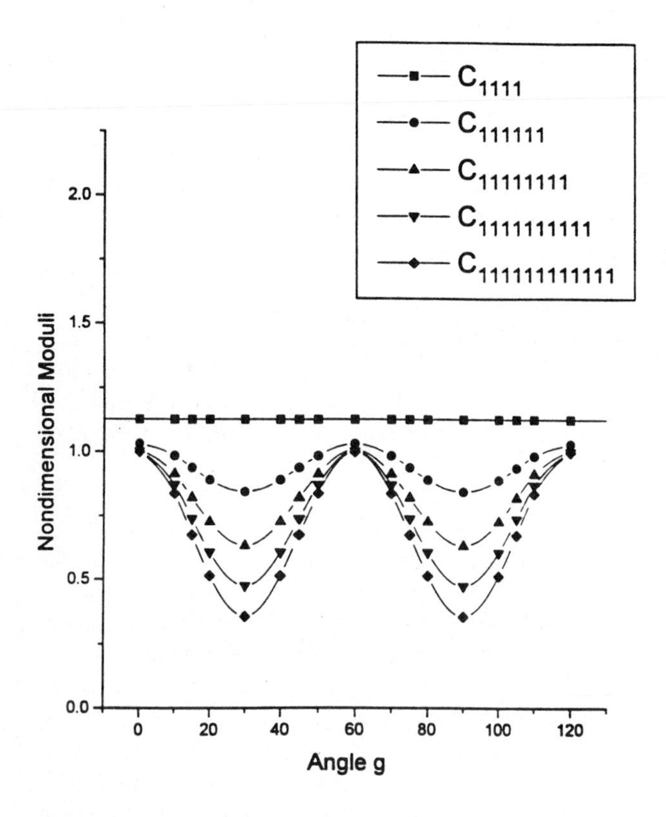

Fig. 5.4. Variations of nondimensional moduli of elasticiy $\bar{C}_{ijk_1\ldots k_\kappa}$ with the rotation of a plane frame of reference by an angle γ

The variation of the moduli with γ is plotted in Fig. 5.4. It is noted that the nonscale macromodulus $\bar{C}_{1111}$, corresponding to $\kappa = 2$, is indeed independent of γ, i.e., isotropic in the plane. On the contrary, the macromoduli $\bar{C}_{ijk_1\ldots k_\kappa}$ for $\kappa = 4, 6, 8, \ldots$ are anisotropic. It can be seen from Fig. 5.4 that all $\bar{C}_{ijk_1\ldots k_\kappa} \to 0$ as $\kappa \to \infty$ for any angle γ except for $\gamma = 0°, 60°, 120°$ where $\bar{C}_{ijk_1\ldots k_\kappa} \to 1$, which are the angles that identify the directions of the doublets (see Fig. 5.2). It may then be concluded that in the first approximation, $\kappa = 2$, eqns. (5.16) model the continuum-like behavior of solids, whereas in the other approximations, $\kappa = 4, 6, \ldots$, eqns. (5.16) also reflect discrete-like features of the solid, in a manner that increases with κ.

In this sense, doublet mechanics may be concluded to be capable of modeling solids in view of their dual and to some extent contradictory discrete-continuous nature. The power of such dual-representation capability is evident in the discussion of isotropy. The basal plane of the cubic-tetrahedral arrangement is isotropic only in the continuum, nonscale approximation. Thus, isotropy is a scale-related notion—a fact that is of course physically evident, as no material may be argued to be isotropic at all dimensional scales, down to its most elementary component level. What is promising for the doublet mechanics approach is the fact that this theory is capable of modeling such observation, and recourse to different theories for different dimensional scales is avoided altogether.

According to eqns. (5.56)–(5.57), the velocity ratio is $C_1 \equiv V_{\text{op}}/V_{\text{os}} = \sqrt{3}$. By comparison with the well-known relation (Fung 1977)

$$C_1 = \sqrt{\frac{2\,(1-\nu)}{1-2\nu}}, \tag{5.68}$$

this leads to Poisson's constant $\nu = 1/4$. This value is conventionally assumed in seismology for the Earth's crust (Leet 1938, Macelwane 1949, Båth 1968). A theoretical validation of this assumption is furnished by doublet mechanics. Any isotropic tensor in accordance with the relation (5.19) must have the form

$$C_{ijkl} = \lambda\,(\delta_{ij}\,\delta_{kl} + \delta_{ik}\,\delta_{jl} + \delta_{il}\,\delta_{jk}). \tag{5.69}$$

This tensor differs from the conventional (continuum) moduli of elasticity in that it is also endowed with the additional symmetry $C_{ijlk} = C_{ikjl}$ which is not necessary in continuum mechanics. In turn, the symmetry imposes the equality of the Lame' constants, i.e., $\lambda = \mu$. This reduces the number of independent moduli to only one. Of course, this conclusion holds only for materials with such microstructure and properties that satisfy the assumptions employed for the derivation of eqn. (5.19). By applying the well-known relationship

$$\nu = \frac{\lambda}{2\,(\lambda + \mu)}. \tag{5.70}$$

it is then concluded that $\nu = 1/4$, which was to be shown.

The developed micromechanical elastodynamics theory is applicable not only to the unbounded isotropic granular solids but to any similar particulate media, i.e., any media with microstructure characterized geometrically by finite particle sizes (or finite central particle distances) η and physically by longitudinal particle interactions (5.3) only.

The term microstructure is relative, in our context, in the sense that the wave dispersion and wave retardation depend on the wavelength λ and the characteristic distance η not separately, but in a nondimensional combination $l \equiv \lambda/\eta$. Thus the internal size η is significant only in comparison with the wavelength λ. This may be arbitrary large and, consequently, the size η may also be arbitrary large provided the ratio l has finite values $l \geq 2$. On these bases, the absolute dimension of the typical internal structure is not a determining factor concerning the applicability of the present theory. To illustrate the point, three classes of materials are here briefly considered, that possess very different internal dimensional scales.

First, consider crystals with typical interatomic spacing $\eta \sim 10^{-8}$cm. The phenomena of wave dispersion and retardation in linear monatomic crystals are well-known in solid state physics where they are modeled by eqn. (5.27). Experimental verifications of eqn. (5.29) are given, for instance, in Kittel (1986) and Cochran (1973).

Equation (5.27) can be generalized to higher dimensions for two- and three-dimensional crystal arrangements. The corresponding theoretical results obtained in solid state physics have been verified experimentally for many spatial crystal structures including the f.c.c. such as aluminum and copper. The f.c.c. structure is equivalent to a regular pyramidal packing H_6 (Deresiewicz 1958, pp.237–238) which has crystallographic planes identical with the basal plane of the cubical-tetrahedral packing H_4 considered in this section.

Second, consider granular materials with the characteristic granule sizes η in the range 10^{-1}–10^2 cm (sand, gravel, rubble, boulders), where the impact of particle sizes on wave propagation has been directly studied in a few experimental works. The earliest of these (Iida 1938) established that the phase velocities of P- and S-waves in dry sand depend on the particle diameter d and slightly rise as d increases. These data first indicated that wave dispersion was associated with size parameters, but was at variance with the phenomenon of wave retardation. The fact that the phase velocities of plane P-waves are sensitive to the granule sizes was also observed in other direct experiments with dry sand (Matsukawa and Hunter 1956).

Several later studies brought indirect experimental data concerning the influence of particle sizes on wave velocities in granular massives. Among these, Trent (1989) dealt with two arrays of 270 like spheres which had diameter $d = 2$ mm and occupied equal volumes. The first array had 480 interparticle bonds, the second one 397, i.e., 21% less. By performing numerical experiments based on the so-called distinct element method (Cundall and Strack 1979), it was established that the phase velocity of P-wave in the second array is 37% less then in the first array (1050 m/s versus 1440 m/s). There is only one difference in the two arrays which is responsible for the change in velocities—the difference in numbers N of the bonds. Meanwhile in any regular n-valence packing H_3 to H_6 (should they have equal volumes) the number N and diameter d are inversely correlated. So a decrease of bond numbers is, in general, equivalent to an increase of particle size. Therefore the retardation of wave observed in the second granular array tested may be attributed either directly to a reduced number of particle bonds or indirectly to an increase of particle diameters d and, consequently, to a decrease of the ratio $l \equiv \lambda/d$, λ being the wavelength. This means that the waves slow down as the scaling parameter l declines, a result that is in agreement with the quality of Fig. 5.3.

Finally, consider the Earth's crust with $\eta \sim 10^6$ cm (= 10 km). As discussed in Granik and Ferrari (1994) the doublet mechanical approach is applicable in the context of plane-wave seismology. In particular, in the above reference, comparisons were made between classical seismic data (Leet 1938, page 261, Miyabe 1935, Pilant 1979, page 254, Fig. 7-1), and scale-accounting predictions of DM that are concerned with the effects of the particulate structure of the crust on the velocities of longitudinal waves. The results showed good agreement (no more than a 5% discrepancy) between the predictions and the three sets of seismic data.

6. Reflection of Plane Waves

M. Zhang and M. Ferrari

6.1 Introduction

Microscopically isotropic, infinite plane regular assemblies of discrete nodes capable of elastic axial interactions were shown in Chap. 5 to sustain both longitudinal and shear vertical plane waves for all values of the dimensionless scale parameter $l = \lambda/\eta$, where λ is the wavelength and η is the internodal distance. Ibidem, it was also shown that the incorporation of scale effects permits the modelling of physical observations that are otherwise intractable in terms of a general theory. Among these are the phenomena of dispersion and retardation of both P- and S-waves, which are incompatible with homogeneous continuum linear elasticity, but are successfully predicted employing the microstructure-accounting methods of multi-scale doublet mechanics (DM).

In Chap. 5, the elastodynamic study was limited to infinite media, and in this chapter we extend that work, by considering the problem of reflection of plane waves at the free surface of macroscopically isotropic solids. While the problem of plane elastodynamics in solids has received considerable attention in the literature (see discussion in Chap. 5), the issue of wave reflection has, to the best of the authors' knowledge, never been addressed.

In the first part of this chapter, the non-scale version of the problem of plane wave reflection is considered, with results that are in compliance with those of continuum linear elastodynamics. However, in later sections it is demonstrated that the simplest scale variable of the theory elicits results that are qualitatively different from the classical ones. In particular, it is found that the critical angles of mode conversion, the phase changes and the amplitude ratios are dependent on the dimensionless scale parameter, and thus on discrete size for a fixed wavelength. The dependence is more pronounced at shorter wavelengths.

The following analysis, for mathematical simplification and physical explicitness, is based on the following assumptions:

(i) no body forces are acting on the discret domain;
(ii) the doublets are capable of axial microstresses only;
(iii) the doublet constitutive response is linear elastic and local, with microstresses in α-th doublet depending on the α-th microstrain only;

(iv) the discret packing is macroscopically isotropic in the plane of propagation.

It is noted that (iv) is satisfied by choosing the plane of propagation to be the basal plane of the cubic-tetrahedral packing, which is elastically isotropic in the nonscale (macroscopic) limit, but becomes anisotropic once scale effects are accounted for (Chap. 5).

In this chapter, the first and fundamental scale-accounting version of the doublet mechanical approach is employed, corresponding to the truncation of the displacement expansion at $M = 2$ in eqn. (1.5). This choice retains the advantage of analytic treatment, while sufficing to establish the above-mentioned qualitative features of the scale-accounting treatment. The presently employed methods of analysis are not confined to the specific packing here discussed, but are applicable to any other discrete array of interest.

6.2 Model and Formulation of Reflection of Incident Waves

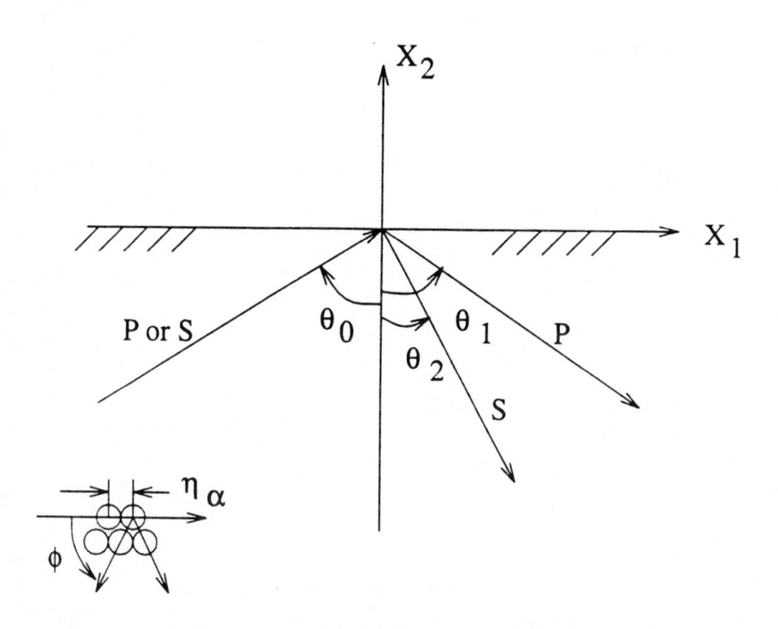

Fig. 6.1. Reflection of an incident P-wave at the interface of discrete medium and free space with a sketch of corresponding doublet orientation (τ_{3i} parallel to x_1 axis); θ_0 is the incident angle of the incident P-wave; θ_1 is the reflection angle of the reflected P-wave; θ_2 is the reflection angle of the reflected S-wave; ϕ is the angle between the directions of two doublets.

We consider single-frequency time-harmonic waves only. Fig. 6.1 depicts a plane P-wave traveling at an incident angle θ_0. Upon reflection with the free surface $x_2 = 0$ this wave gives rise to a reflected P-wave and a reflected S-wave with reflection angles θ_1 and θ_2, respectively.

The incident and reflected waves propagating in the half-space $x_2 < 0$ are defined by the following equations

$$\mathbf{u}^{(n)} = A_n \mathbf{d}^{(n)} \exp[i\eta^{(n)}] \tag{6.1}$$

where

$$\eta^{(n)} = k_n(\mathbf{x} \cdot \mathbf{p}^{(n)} - c_n t) \tag{6.2}$$

where A_n are the amplitudes of the waves and $\mathbf{d}^{(n)}$ are the unit vectors of particle motion; $\mathbf{x}$ is the position vector and $\mathbf{p}^{(n)}$, the unit vectors of propagation; k_n are the wave numbers and c_n are the phase velocities of wave propagation. The index n is assigned the value of 0 for the incident P-wave, 1 for the reflected P-wave and 2 for the reflected S-wave.

From the geometry of Fig. 7.1 we have, for the incident P-wave

$$\mathbf{d}^{(0)} = \mathbf{p}^{(0)}, \quad c_0 = c_{\mathrm{L}} \tag{6.3}$$

$$\mathbf{p}^{(0)} = \sin\theta_0 \mathbf{i}_1 + \cos\theta_0 \mathbf{i}_2 \tag{6.4}$$

$$\{\mathbf{u}^{(0)}\} = \left\{ \begin{array}{c} u_1^{(0)} \\ u_2^{(0)} \end{array} \right\} \tag{6.5}$$

$$= \left\{ \begin{array}{c} A_0 \sin\theta_0 \exp[ik_0(x_1 \sin\theta_0 + x_2 \cos\theta_0 - c_0 t)] \\ A_0 \cos\theta_0 \exp[ik_0(x_1 \sin\theta_0 + x_2 \cos\theta_0 - c_0 t)] \end{array} \right\} \tag{6.6}$$

for an incident S-wave

$$\mathbf{d}^{(0)} = \mathbf{i}_3 \times \mathbf{p}^0, \quad c_0 = c_{\mathrm{T}} \tag{6.7}$$

$$\mathbf{p}^{(0)} = \sin\theta_0 \mathbf{i}_1 + \cos\theta_0 \mathbf{i}_2 \tag{6.8}$$

$$\mathbf{d}^{(0)} = -\cos\theta_0 \mathbf{i}_1 + \sin\theta_0 \mathbf{i}_2 \tag{6.9}$$

$$\{\mathbf{u}^{(0)}\} = \left\{ \begin{array}{c} u_1^{(0)} \\ u_2^{(0)} \end{array} \right\} \tag{6.10}$$

$$= \left\{ \begin{array}{c} -A_0 \cos\theta_0 \exp[ik_0(x_1 \sin\theta_0 + x_0 \cos\theta_0 - c_0 t)] \\ A_0 \sin\theta_0 \exp[ik_0(x_1 \sin\theta_0 + x_2 \cos\theta_0 - c_0 t)] \end{array} \right\} \tag{6.11}$$

for the reflected P-wave

$$\mathbf{p}^{(1)} = \sin\theta_1 \mathbf{i}_1 - \cos\theta_1 \mathbf{i}_2 \tag{6.12}$$

$$\mathbf{d}^{(1)} = \mathbf{p}^{(1)}, \quad c_1 = c_{\mathrm{L}} \tag{6.13}$$

$$\{\mathbf{u}^{(1)}\} = \left\{ \begin{array}{c} u_1^{(1)} \\ u_2^{(1)} \end{array} \right\} \tag{6.14}$$

$$= \left\{ \begin{array}{c} A_1 \sin\theta_1 \exp[ik_1(x_1 \sin\theta_1 - x_2 \cos\theta_1 - c_1 t)] \\ -A_1 \cos\theta_1 \exp[ik_1(x_1 \sin\theta_1 - x_2 \cos\theta_1 - c_1 t)] \end{array} \right\} \tag{6.15}$$

and finally for the reflected S-wave

$$\mathbf{p}^{(2)} = \sin\theta_2 \mathbf{i}_1 - \cos\theta_2 \mathbf{i}_2 \tag{6.16}$$

$$\mathbf{d}^{(2)} = \mathbf{i}_3 \times \mathbf{p}^{(2)}, \quad c_2 = c_{\mathrm{T}} \tag{6.17}$$

$$\mathbf{d}^{(2)} = \cos\theta_2 \mathbf{i}_1 + \sin\theta_2 \mathbf{i}_2 \tag{6.18}$$

$$\{\mathbf{u}^{(2)}\} = \left\{ \begin{array}{c} u_1^{(2)} \\ u_2^{(2)} \end{array} \right\} \tag{6.19}$$

$$= \left\{ \begin{array}{c} A_2 \cos\theta_2 \exp[ik_2(x_1 \sin\theta_2 - x_2 \cos\theta_2 - c_2 t)] \\ A_2 \sin\theta_2 \exp[ik_2(x_1 \sin\theta_2 - x_2 \cos\theta_2 - c_2 t)] \end{array} \right\} \tag{6.20}$$

In the above, c_{L} and c_{T} are the propagation velocities of P- and S-waves respectively. The derivatives of the displacement of the incident P-wave are

$$\begin{aligned}
\frac{\partial u_1^{(0)}}{\partial x_1} &= iA_0 k_0 \sin^2\theta_0 \exp[i\eta^{(0)}] \\
\frac{\partial u_1^{(0)}}{\partial x_2} &= iA_0 k_0 \sin\theta_0 \cos\theta_0 \exp[i\eta^{(0)}] \\
\frac{\partial u_2^{(0)}}{\partial x_1} &= iA_0 k_0 \cos\theta_0 \sin\theta_0 \exp[i\eta^{(0)}] \\
\frac{\partial u_2^{(0)}}{\partial x_2} &= iA_0 k_0 \cos^2\theta_0 \exp[i\eta^{(0)}] \\
\frac{\partial^2 u_1^{(0)}}{\partial x_1^2} &= -A_0 k_0^2 \sin^3\theta_0 \exp[i\eta^{(0)}] \\
\frac{\partial^2 u_1^{(0)}}{\partial x_2^2} &= -A_0 k_0^2 \sin\theta_0 \cos^2\theta_0 \exp[i\eta^{(0)}] \\
\frac{\partial^2 u_2^{(0)}}{\partial x_1^2} &= -A_0 k_0^2 \sin^2\theta_0 \cos\theta_0 \exp[i\eta^{(0)}] \\
\frac{\partial^2 u_2^{(0)}}{\partial x_2^2} &= -A_0 k_0^2 \cos^3\theta_0 \exp[i\eta^{(0)}] \\
\frac{\partial^2 u_2^{(0)}}{\partial x_1 \partial x_2} &= -A_0 k_0^2 \sin\theta_0 \cos^2\theta_0 \exp[i\eta^{(0)}] \\
\frac{\partial^2 u_1^{(0)}}{\partial x_1 \partial x_2} &= -A_0 k_0^2 \sin^2\theta_0 \cos\theta_0 \exp[i\eta^{(0)}].
\end{aligned} \tag{6.21}$$

Derivatives for the other waves are similarly defined.

Via eqn. (1.40) the free stress boundary conditions at plane $x_2 = 0$ may be expressed as

$$\sum_n \sigma_{21}^{(n)} = 0 \tag{6.22}$$

$$\sum_n \sigma_{22}^{(n)} = 0 \tag{6.23}$$

where the summation is performed over each wave n, and the superscript M is dropped for notational conventional convenience.

In analogy with continuum elastodynamics (Achenbach 1980), for a given incident wave, the amplitudes, the unit propagation vectors, and the wavenumber must be computed from the boundary conditions. Thus, the problem is reduced to solving for the amplitude coefficients and reflection angles of reflected waves, on the basis of the above stated stress free boundary conditions. This approach also allows us to develop families of solutions to the reflection of incident waves for various boundary conditions. Some scale and nonscale reflection problems are studied next.

6.3 Reflection of an Incident P-Wave: Non-scale Analysis

6.3.1 Doublet Axis Parallel to the Reflecting Surface

According to the coordinate system defined in Fig. 6.1, we can write the direction matrix of doublets as follows

$$\begin{bmatrix} \tau_{11} & \tau_{12} \\ \tau_{21} & \tau_{22} \\ \tau_{31} & \tau_{32} \end{bmatrix} = \begin{bmatrix} -\cos\phi & -\sin\phi \\ \cos\phi & -\sin\phi \\ 1 & 0 \end{bmatrix} \tag{6.24}$$

where ϕ is 60° for the basal plane of the cubic tetrahedral packing. For the nonscale case ($M = 1$) this plane is elastically isotropic (see Chap. 5). The stress relation (1.40) for $M = 1$ is thus reduced to

$$\sigma_{ji} = \sum_{\alpha=1}^{3} \tau_{\alpha i}\, \tau_{\alpha j}\, p_\alpha \tag{6.25}$$

where, for plane waves, $i, j = 1, 2$. From the boundary conditions (6.22) and (6.23), constitutive relation (5.3), and kinematic equation (1.12), we obtain the following relations between the incident and reflected waves

$$\begin{aligned} &A_0 k_0 [\sin^2\theta_0 + 3\cos^2\theta_0] \exp[i\eta^{(0)}] \\ &\quad + A_1 k_1 [\sin^2\theta_1 + 3\cos^2\theta_1] \exp[i\eta^{(0)}] \\ &\quad - A_2 k_2 \sin(2\theta_2) \exp[i\eta^{(2)}] = 0 \end{aligned} \tag{6.26}$$

$$A_0 k_0 \sin 2\theta_0 \exp[i\eta^{(0)}] - A_1 k_1 \sin 2\theta_1 \exp[i\eta^{(1)}] - A_2 k_2 \cos(2\theta_2) \exp[i\eta^{(2)}] = 0 \tag{6.27}$$

Since eqns. (6.26) and (6.27) are valid for all values of x_1 and t at $x_2 = 0$, the existence of solutions of the set of equations requires that the exponential must appear as factors in both equations (Achenbach 1980). This will be satisfied only when

$$\eta^0 = \eta^1 = \eta^2 \tag{6.28}$$

We conclude, from inspection of the definition of $\eta^{(n)}$ in eqn. (6.2), that

$$k_0 \sin \theta_0 = k_1 \sin \theta_1 = k_2 \sin \theta_2 = \kappa \tag{6.29}$$

$$k_0 c_{\mathrm{L}} = k_1 c_{\mathrm{L}} = k_2 c_{\mathrm{T}} = \omega \tag{6.30}$$

From eqns. (6.29) and (6.30), it follows that

$$\theta_0 = \theta_1, \quad \theta_2 = \sin^{-1}(\kappa^{-1} \sin \theta_0) \tag{6.31}$$

$$k_1 = k_0, \quad \frac{k_2}{k_0} = \frac{c_{\mathrm{L}}}{c_{\mathrm{T}}} = \kappa \tag{6.32}$$

By employing eqns. (6.31) and (6.32), the algebraic equations for the amplitude ratios A_1/A_0 and A_2/A_0 are obtained from eqns. (6.26) and (6.27):

$$\frac{A_1}{A_0}(\sin^2 \theta_0 + 3 \cos^2 \theta_0) - \frac{A_2}{A_0} \kappa \sin 2\theta_2 = -(\sin^2 \theta_0 + 3 \cos^2 \theta_0) \tag{6.33}$$

$$\frac{A_1}{A_0} \sin 2\theta_0 + \frac{A_2}{A_0} \kappa \cos 2\theta_2 = \sin 2\theta_0 \tag{6.34}$$

A typical plot of amplitude ratios A_1/A_0, A_2/A_0 versus the angle of incidence is shown in Fig. 6.2 for the standard value of material constant $\kappa = \sqrt{3}$ (Kolsky 1963). This value of the material constant was chosen for ease of comparison with literature results. It is found that the results obtained using DM theory in the first degree approximation ($M = 1$, nonscale) corresponds exactly to those of classical continuum elasticity (Kolsky 1963, Achenbach 1980).

It is observed from Fig. 6.2 that:

(i) for normal incidence ($\theta_0 = 0$), the amplitude of the reflected S-wave is zero, the incident P-wave is reflected as a P-wave only, and the amplitude of the wave is equal to that of the incident wave with a phase change of π;

(ii) for an angle of incidence about 45°, the amplitude of the reflected S wave reaches the maximum, which is greater than that of the incident wave;

(iii) for the angles of incidence $\theta_0 = 60°$ and about 80° , the incident P-wave is reflected as a S-wave only, which is known as mode conversion phenomena;

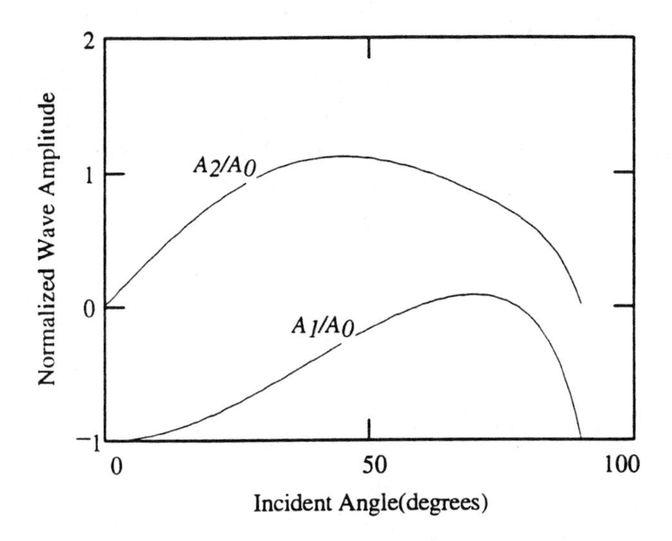

Fig. 6.2. Amplitude ratios of the reflected waves to the incident wave for an incident P-wave in the nonscale case of M=1. A_0 is the amplitude of the incident P-wave; A_1 is the amplitude of the reflected P-wave; A_2 is the amplitude of the reflected S-wave.

(iv) for $\theta_0 = 90°$, the reflected S wave vanishes and the incident P-wave is again reflected as a P-wave;

(v) the reflected P-wave have same phase as the incident P-wave between about 60°–80°, and otherwise it is 180° out of phase to the incident wave.

6.3.2 Doublet Axis at an Angle γ with Respect to the Reflecting Surface

The geometry of this case is sketched in Fig. 6.3, and the direction matrix of the rotation may be written as

$$[\mathring{\tau}_{ij}] = \begin{bmatrix} \mathring{\tau}_{11} & \mathring{\tau}_{12} \\ \mathring{\tau}_{21} & \mathring{\tau}_{22} \\ \mathring{\tau}_{31} & \mathring{\tau}_{32} \end{bmatrix} = \begin{bmatrix} -\cos(\phi-\gamma) & -\sin(\phi-\gamma) \\ \cos(\phi+\gamma) & -\sin(\phi+\gamma) \\ \cos\gamma & -\sin\gamma \end{bmatrix} \tag{6.35}$$

By eqn. (1.12) and eqn. (5.3), the stress relation (1.40) may be expressed in terms of the components of displacement as

$$\sigma_{21}^{(n)} = \sum_n \sum_{\alpha=1}^{3} \left[\mathring{\tau}_{\alpha 2}(\mathring{\tau}_{\alpha 1})^3 \frac{\partial u_1^{(n)}}{\partial x_1} + (\mathring{\tau}_{\alpha 1}\mathring{\tau}_{\alpha 2})^2 \left(\frac{\partial u_2^{(n)}}{\partial x_1} + \frac{\partial u_1^{(n)}}{\partial x_2} \right) + \mathring{\tau}_{\alpha 1}(\mathring{\tau}_{\alpha 2})^3 \frac{\partial u_2^{(n)}}{\partial x_2} \right] \tag{6.36}$$

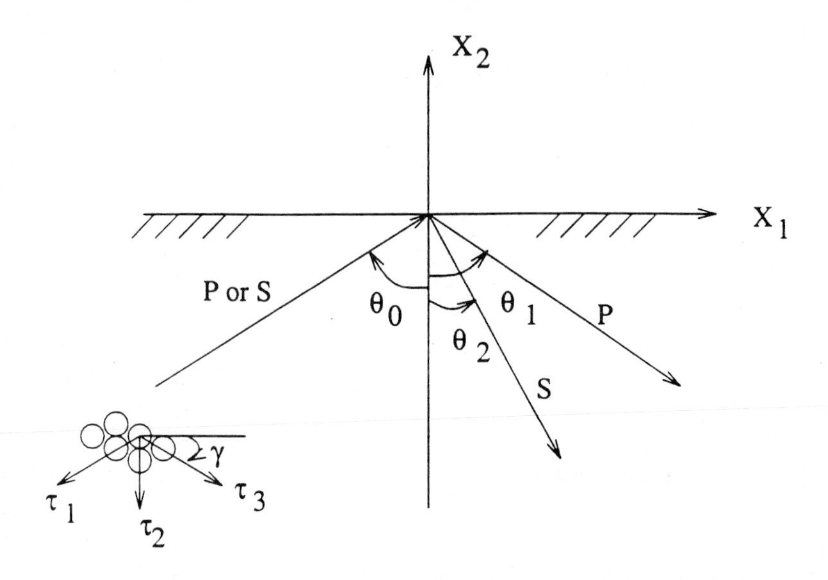

Fig. 6.3. Reflection of an incident P-wave at the interface of discrete medium and free space with the sketch of corresponding doublet orientation (τ_3 in an angle of γ with respect to x_1 axis); θ_0 is the incident angle of the incident P-wave; θ_1 is the reflection angle of the reflected P-wave; θ_2 is the reflection angle of the reflected S-wave.

$$\begin{aligned}\sigma_{22}^{(n)} &= \sum_n \sum_{\alpha=1}^{3} \left[\overset{\circ}{\tau}_{\alpha 2} (\overset{\circ}{\tau}_{\alpha 1})^3 \frac{\partial u_1^{(n)}}{\partial x_1} + \overset{\circ}{\tau}_{\alpha 1} (\overset{\circ}{\tau}_{\alpha 2})^3 \left(\frac{\partial u_1^{(n)}}{\partial x_2} + \frac{\partial u_2^{(n)}}{\partial x_1} \right) \right. \\ &\quad \left. + (\overset{\circ}{\tau}_{\alpha 2})^4 \frac{\partial u_2^{(n)}}{\partial x_2} \right] \end{aligned} \tag{6.37}$$

Substituting eqns. (6.36) and (6.37) into eqns. (6.22) and (6.23), and following the same procedure as in Sect. 6.3.1, it is proven that the angle of the reflected P-wave is equal to the angle of the incident P-wave and the angle of the reflected S-wave still satisfies eqn. (6.31). The amplitude ratio expressions can thus be reduced to

$$\begin{aligned}&\sum_\alpha \overset{\circ}{\tau}_{\alpha 2} (\overset{\circ}{\tau}_{\alpha 1})^3 (A_0 k_0 \sin^2 \theta_0 + A_1 k_1 \sin^2 \theta_1 + A_2 k_2 \cos \theta_2 \sin \theta_2) \\ &\quad + \sum_\alpha (\overset{\circ}{\tau}_{\alpha 1})^2 (\overset{\circ}{\tau}_{\alpha 2})^2 (A_0 k_0 \sin 2\theta_0 - A_1 k_1 \sin 2\theta_1 \\ &\qquad - A_2 k_2 \cos^2 \theta_2 + A_2 k_2 \sin^2 \theta_2) \\ &\quad + \sum_\alpha (\overset{\circ}{\tau}_{\alpha 2})^3 \overset{\circ}{\tau}_{\alpha 1} \Big(A_0 k_0 \cos^2 \theta_0 + A_1 k_1 \cos^2 \theta_1 \\ &\quad - \frac{1}{2} A_2 k_2 \sin 2\theta_2 \Big) = 0 \end{aligned} \tag{6.38}$$

and

$$\sum_{\alpha}(\overset{\circ}{\tau}_{\alpha 2})^2(\overset{\circ}{\tau}_{\alpha 1})^2(A_0 k_0 \sin^2\theta_0 + A_1 k_1 \sin^2\theta_1 + \frac{1}{2}A_2 k_2 \sin 2\theta_2)$$
$$+\sum_{\alpha}\overset{\circ}{\tau}_{\alpha 1}(\overset{\circ}{\tau}_{\alpha 2})^3(A_0 k_0 \sin 2\theta_0 - A_1 k_1 \sin 2\theta_1$$
$$- A_2 k_2 \cos^2\theta_2 + A_2 k_2 \sin^2\theta_2)$$
$$+\sum_{\alpha}(\overset{\circ}{\tau}_{\alpha 2})^4(A_0 k_0 \cos^2\theta_0 + A_1 k_1 \cos^2\theta_1 - \frac{1}{2}A_2 k_2 \sin 2\theta_2) = 0. \quad (6.39)$$

The ratios of the amplitudes of the reflected waves to the amplitude of the incident P-wave have been plotted for varying angle of γ and it was found that the reflection coefficients are independent of the rotation angle of γ and the dependence of amplitudes of reflected waves on the incident angles is exactly the same as in Sect. 6.3.1.

6.4 Reflection of an Incident P-Wave: Scale Analysis

The first order scale-accounting case is addressed next by choosing the order of approximation $M = 2$. Referring to eqns. (1.12) and (1.40), the stress relations may be reduced to

$$\sigma_{ji} = A_\circ \sum_{\alpha} \overset{\circ}{\tau}_{\alpha j}\left(\overset{\circ}{\tau}_{\alpha i}\epsilon_\alpha - \frac{\eta_\alpha}{2}\overset{\circ}{\tau}_{\alpha k}\overset{\circ}{\tau}_{\alpha i}\frac{\partial \epsilon_\alpha}{\partial x_k}\right) \quad (6.40)$$

In this case, unlike the case of $M = 1$ in which the stresses are the function of strain only, the stresses are the function of both strains and the derivatives of strains. Following the same steps as the Case I, by employing the stress-strain relation (6.40) and the stress free boundary condition (6.22) and (6.23), the following relations are derived

$$\theta_1 = \theta_0, \quad k_1 = k_0 \quad (6.41)$$

$$\theta_2 = \sin^{-1}(\kappa^{-1}\sin\theta_0) \quad (6.42)$$

where $\kappa = k_2/k_0$.

With eqns. (6.41) and (6.42) and the stress free boundary conditions (6.22) and (6.23), the algebraic equations for the amplitude ratios of the incident and the reflected waves, A_1/A_0 and A_2/A_0, can be expressed as follows,

$$\left[\frac{3}{4}\sin\theta_0\cos\theta_0 - i\frac{\sqrt{3}}{16}\frac{\pi}{l}\sin\theta_0 - i\frac{11}{16}\sqrt{3}\frac{\pi}{l}\sin\theta_0\cos^2\theta_0\right]\frac{A_1}{A_0}$$
$$+\left[\frac{\sqrt{3}}{8} + \frac{\sqrt{3}}{4}\cos^2\theta_0 + i\frac{5}{16}\frac{\pi}{l}\sqrt{2+\cos^2\theta_0}\right.$$

$$- i\frac{5}{16}\frac{\pi}{l}\sqrt{2+\cos^2\theta_0}\cos^2\theta_0\Bigg]\frac{A_2}{A_0}$$
$$= \frac{3}{4}\sin\theta_0\cos\theta_0 + i\frac{\sqrt{3}}{16}\frac{\pi}{l}\sin\theta_0 + i\frac{\sqrt{3}}{2}\frac{\pi}{l}\sin\theta_0\cos^2\theta_0, \tag{6.43}$$

$$\left[-\frac{3}{8} - \frac{3}{4}\cos^2\theta_0 + i\frac{9}{16}\sqrt{3}\frac{\pi}{l}\cos\theta_0\right]\frac{A_1}{A_0}$$
$$+ \left[\frac{1}{4}\sqrt{6+3\cos^2\theta_0}\sin\theta_0 - i\frac{9}{16}\frac{\pi}{l}\sin\theta_0\right]\frac{A_2}{A_0}$$
$$= \frac{3}{8} + \frac{3}{4}\cos^2\theta_0 + i\frac{9}{16}\frac{\pi}{l}\sqrt{3}\cos\theta_0. \tag{6.44}$$

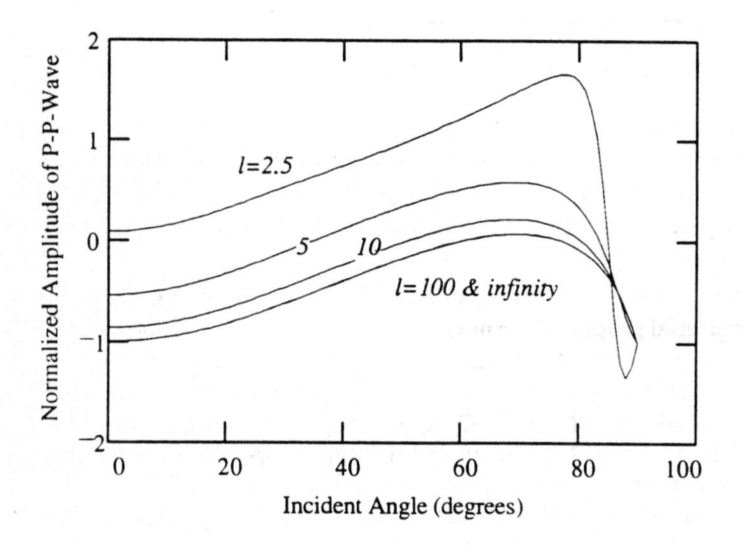

Fig. 6.4. Real components of amplitude ratios of the reflected P-wave to the incident P-wave in the scale case of $M=2$ with l, the dimensionless scale factor as a parameter.

The ratios A_1/A_0 and A_2/A_0 versus incident angles are plotted in Figs. 6.4 and 6.5 for different values of scale factors $l = \lambda/\eta_\alpha$, where λ is the wave length and η_α is the central distance of doublets as shown in Fig. 6.1. For $l = \infty$, the nonscale results of Sect. 6.3.1 are retrieved. In the figures throughout this chapter, the following nomenclature is employed for the axis labels: the letter(s) before the dash line indicates the types of incident wave and the letter(s) after the dash line indicates the types of reflected waves. For instance, a reflected P-wave due to the incident S-wave is identified as an S-P-wave.

From Figs. 6.4 and 6.5, it is observed that:

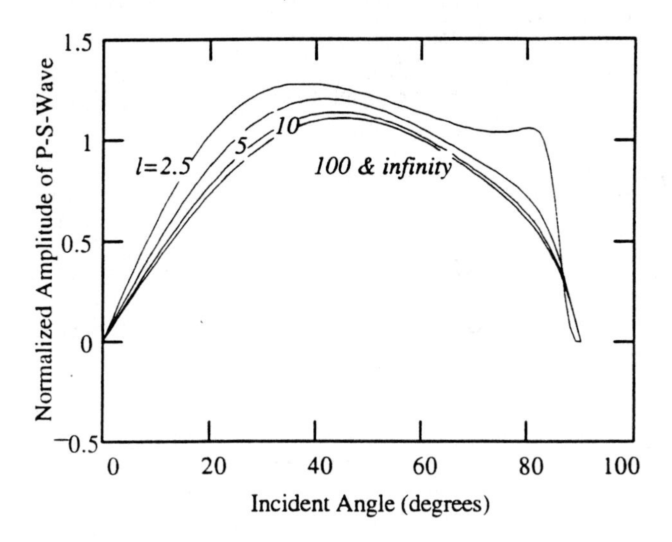

Fig. 6.5. Real components of amplitude ratios of the reflected S-wave to the incident P-wave in the scale case of M=2 with l, the dimensionless scale factor as a parameter.

(i) the amplitudes of the reflected waves not only depend on the incident angle and material property κ as in the nonscale case, but also depend on the scale factor l, which reflects the size of the doublet; the dependence is especially strong at wavelengths comparable to the particle size.
(ii) For incident angle $\theta_0 = 0$, the reflected S-wave vanishes and incident P-wave is reflected as a P-wave; the amplitude of the reflected P-wave decreases sharply with l.
(iii) At grazing incidence ($\theta_0 = 90°$) no S-wave is reflected and A_1/A_0, which is independent of scale factors at this angle, become unity.
(iv) Similar to the nonscale case, mode conversion occurs at two angles, but the values of the angles vary with the scale factor l.
(v) the amplitudes of the reflected S-waves increase appreciably with the decrease of l.

The curves for $l = 2.5$ in Figs. 6.4 and 6.5 exhibit features that differ qualitatively from the other curves in the same figures, such as the intersection with curves corresponding to higher values of the dimensionless scale parameters l. It is however noted that the plots correspond to $l = 2.5$, a value outside the valid range of $5 < l < \infty$ recommended for the second degree approximation (Granik and Ferrari 1993), should thus be considered to be poor approximations. For such scale ranges, use should be made of scale approaches involving $M > 2$, following exactly the method of analysis presented above. Still, it is of interest to retain the case $l = 2.5$ in our study, in that different quantities are approximated with different degrees of precision for the same values of 1 and M. An example of this will be shown later.

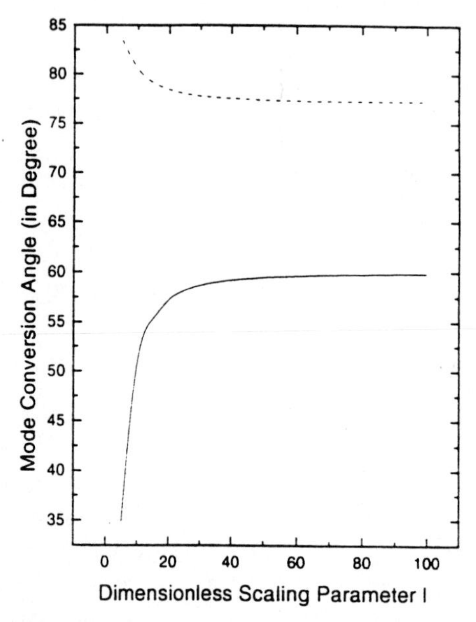

Fig. 6.6. Two mode conversion angles for an incident P-wave versus the dimensionless scale factor l; — the first mode conversion angle, $\cdots$ the second mode conversion angle.

The two mode conversion angles for an incident P-wave are plotted in Fig. 6.6 against the dimensionless scale factor l. In Fig. 6.6 and 6.11, the value of l is taken within the range $(5 < l < \infty)$ that DM theory $(M = 2)$ applies. From Fig. 6.6 it is observed that the first critical angle of mode conversion increases sharply with l when $l < 30$ and changes slowly when $30 < l < 100$, and approaches the angle 60° of the nonscale case when $l > 100$. The second critical angle of mode conversion decreases moderately with the increase of l and approaches the angle 77° of the nonscale case when $l > 100$. It is shown that the results of nonscale case are recovered at $l \simeq 100$.

The phase changes of the reflected P-wave and S-wave with respect to that of the incident P-wave are plotted in Figs. 6.7 and 6.8 against the angle of incidence with l as a parameter. The phase under consideration is defined as

$$\phi = \tan^{-1} \frac{\mathrm{Im}(A_n/A_0)}{\mathrm{Re}(A_n/A_0)} \tag{6.45}$$

where $\mathrm{Im}(A_n/A_0)$ indicates the imaginary component and $\mathrm{Re}(A_n/A_0)$ the real component of A_n/A_0 with $n = 1$ for the reflected P-wave and $n = 2$ for the reflected S-wave.

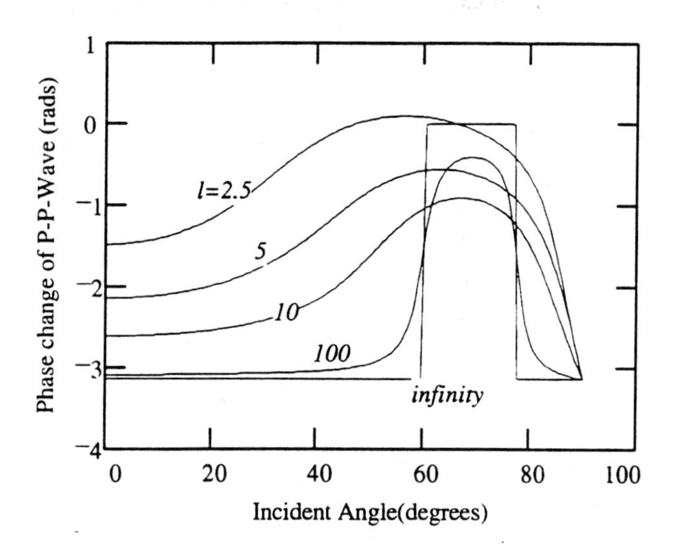

Fig. 6.7. Phase changes of the reflected P-wave with respect to the incident P-wave in the scale case of $M=2$ with l, the dimensionless scale factor as a parameter

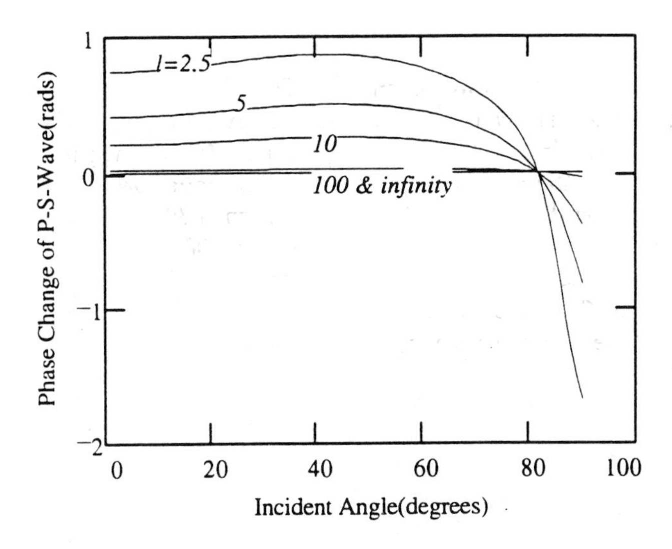

Fig. 6.8. Phase changes of the reflected S-wave with respect to the incident P-wave in the scale case of $M=2$ with l, the dimensionless scale factor as a parameter.

The results of Fig. 6.7 and Fig. 6.8 show that:

(i) For a given wavelength, the particle size, as reflected by the scale factor, has a significant influence on the phase change.
(ii) The phase change for $l = 100$ is very close to that of $l = \infty$, which indicates that for the value of l greater than 100, the results of DM theory are identical with those of classical theory; the phase changes for the values of l less than 100, unlike the nonscale case where the phase difference is either 0 or π, change continuously with the incident angle and increase with the decrease of l.
(iii) The reflected P-wave has abrupt phase changes of π at about 60° and 80° corresponding to the phenomenon of mode conversion of the reflected P-wave shown in Fig. 6.4.
(iv) There is no phase difference between the reflected S-wave and the incident P-wave for the nonscale case for the whole range of incident angles. For the scale case there is a phase difference between them, and the difference increases with the decrease of scale factors l, indicating a departure from the continuum approximation assumed in classical theory. It is noted that the results corresponding to $l = 2.5$ are qualitatively similar to those corresponding to larger values of l, for what pertains to the phase change diagrams.

6.5 Reflection of an Incident S-Wave: Scale Analysis

For an incident S-wave, there is a S-wave and a P-wave reflected at the stress free boundary (Goodier and Bishop 1951). For this case we assign indices $n = 0$ to the incident S wave, $n = 1$ to the reflected P-wave and $n = 2$ to the reflected S-wave.

By strictly following the analogous procedure of the previous sections, we obtain

$$k_0 \sin\theta_0 = k_1 \sin\theta_1 = k_2 \sin\theta_2 \tag{6.46}$$

$$k_0 c_T = k_1 c_L = k_2 c_T \tag{6.47}$$

Similarly, we can conclude from eqns. (6.46) and (6.47) that

$$k_2 = k_0, \quad \theta_2 = \theta_0 \tag{6.48}$$

$$\frac{k_1}{k_0} = \frac{c_T}{c_L} = \kappa^{-1} \tag{6.49}$$

$$\sin\theta_1 = \kappa \sin\theta_0 \tag{6.50}$$

Though we will follow the same procedure to discuss the reflection of an incident S-wave, it is worthwhile to notice at this point that there are significant differences between the reflection of S-waves and P-waves. Since the velocity of propagation for the reflected P-wave is greater than that of the incident S-wave, $\kappa = c_L/c_T$ is always greater than 1; and therefore according to eqn. (6.31), the angle of the reflected S-wave is always less than that of the incident P-wave. By inspection of eqn. (6.50), however, the angle of the reflected P-wave is found to be always greater than that of the incident S-wave. Consequently, there will be a critical angle of incidence at which the reflection angle of P-wave equals $1/(2\pi)$. For those incident angles being greater than the critical angle, a reflected surface wave will be generated and decays exponentially with distance from the free surface. The discussion of surface waves is beyond the scope of this study. In what follows, the reflection of incident S-wave is studied, only for waves with the incident angles which do not exceed the critical angle.

By eqns. (6.48), (6.49) and (6.50) and the stress free boundary conditions (6.22) and (6.23), the stress relation (6.40) for $\kappa = \sqrt{3}$ can be reduced to

$$\begin{aligned}&\Big[-i\frac{9}{2}\frac{\pi}{l}\sqrt{3}\sin\theta_0 + i\frac{27}{4}\frac{\pi}{l}\sqrt{3}\sin\theta_0\cos^2\theta_0 \\ &\quad + \frac{3}{4}\sin\theta_0\sqrt{-2+3\cos^2\theta_0} + i\frac{9}{16}\frac{\pi}{l}\sin\theta_0 - i\frac{15}{16}\frac{\pi}{l}\sin\theta_0\cos^2\theta_0\Big]\frac{A_1}{A_0} \\ &\quad + \left[\frac{3}{4}\cos^2\theta_0 - \frac{3}{8} + i\frac{5}{16}\frac{\pi}{l}\cos\theta_0 - i\frac{5}{16}\frac{\pi}{l}\cos^3\theta_0\right]\frac{A_2}{A_0} \\ &\quad = -\left[i\frac{l}{2}\frac{\pi}{l}\sqrt{3}\cos^3\theta_0 + \frac{3}{4}\cos^2\theta_0 - \frac{3}{8} - i\frac{5}{16}\frac{\pi}{l}\sqrt{3}\cos\theta_0\right]\end{aligned} \tag{6.51}$$

$$\begin{aligned}&\left[\frac{3}{8}\sqrt{3} + i\frac{3}{16}\frac{\pi}{l}\sqrt{3(-2+3\cos^2\theta_0)} - \frac{3}{4}\sqrt{3}\cos^2\theta_0\right]\frac{A_1}{A_0} \\ &\quad + \left[-i\frac{3}{16}\frac{\pi}{l}\sqrt{3}\sin\theta_0 + \frac{3}{4}\cos\theta_0\sin\theta_0\right]\frac{A_2}{A_0} \\ &\quad = i\frac{3}{16}\frac{\pi}{l}\sqrt{3}\sin\theta_0 + \frac{3}{4}\sin\theta_0\cos\theta_0\end{aligned} \tag{6.52}$$

From these equations, A_1/A_0 and A_2/A_0 versus the angle of incidence with scale factor l as a parameter are plotted in Fig. 6.9 and Fig. 6.10. It is found that

(i) there is a critical angle of approximately 35°, which is independent of the scale factor l and consistent with that of eqn. (6.50);
(ii) at $\theta_0 = 0$, the reflected P-wave vanishes, the incident S-wave is reflected as a S-wave only and the magnitudes of reflected S-waves are same as that of incident S-wave for all values of l;
(iii) mode conversion also occurs at two angles, as in the case of incident P-waves, at which the S-wave vanishes and incident S-wave is reflected as P-wave only;

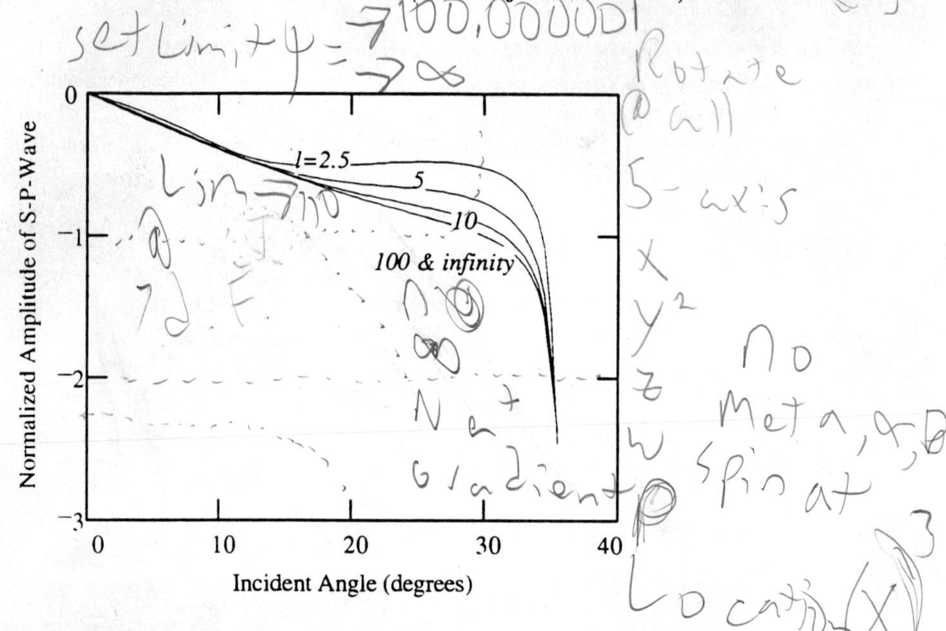

Fig. 6.9. Real components of amplitude ratios of the reflected P-wave to the incident S-wave in the scale case of M=2 with l, the dimensionless scale factor as a parameter.

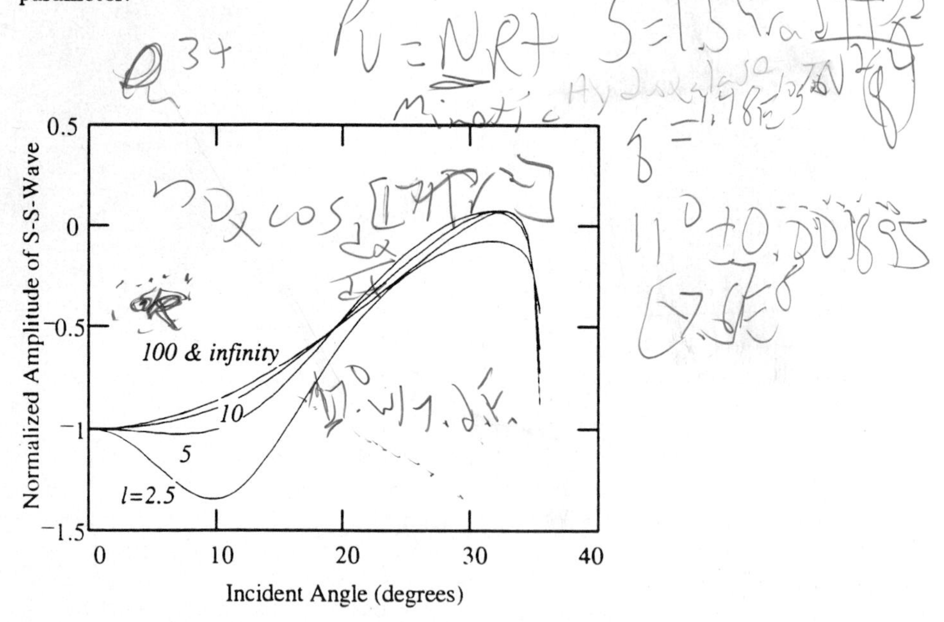

Fig. 6.10. Real components of amplitude ratios of the reflected S-wave to the incident S-wave in the scale case of M=2 with l, the dimensionless scale factor as a parameter.

(iv) the amplitude of the reflected P-wave increases appreciably near the critical angle and reaches a maximum at the critical angle (twice that of the incident S-wave).

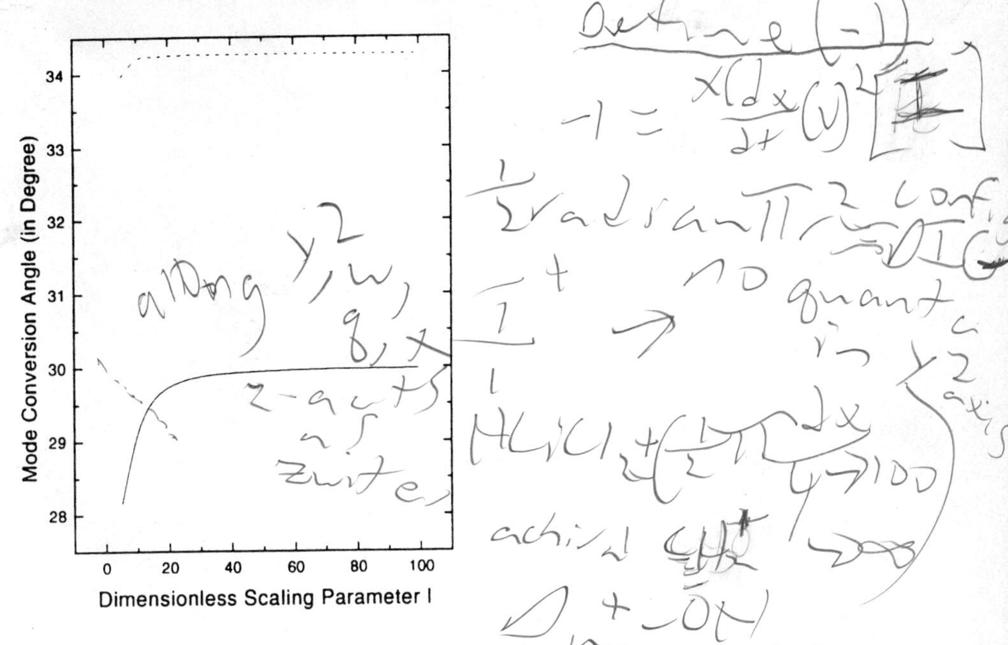

Fig. 6.11. Two mode conversion angles for an incident S-wave versus the dimensionless scale factor l; — the first mode conversion angle, $\cdots$ the second mode conversion angle.

For ease of viewing the dependence of the mode conversion on the microstructural variables, the two mode conversion angle versus the dimensionless scale parameter l, for an incident S-wave, are plotted in Fig. 6.11. It is observed that both critical angles increase sharply with l when $l < 30$, vary slowly when $30 < l < 100$, and reach a maximum value when $l = 100$. The classical case (nonscale) is again retrieved at l greater than 100.

By employing equation (6.45), A_0 denoting the amplitude of an incident S-wave, the phase changes of the reflected waves to an incident S-wave for $M = 2$ and $\kappa = \sqrt{3}$ are plotted in Figs. 6.12 and 6.13. It is observed that

(i) phase changes of reflected waves to an incident S-wave are strongly influenced by scale factor l.

(ii) the reflected P-waves for $l \geq 100$ always have a phase difference of π with respect to that of the incident S-wave; the phase differences between the reflected P-wave and the incident S-wave for other values of l increase with decreasing l and change continuously with the incident angle.

(iii) the reflected S-waves at l greater than 100 also have a phase difference of π with respect to that of incident S-wave and have an abrupt phase

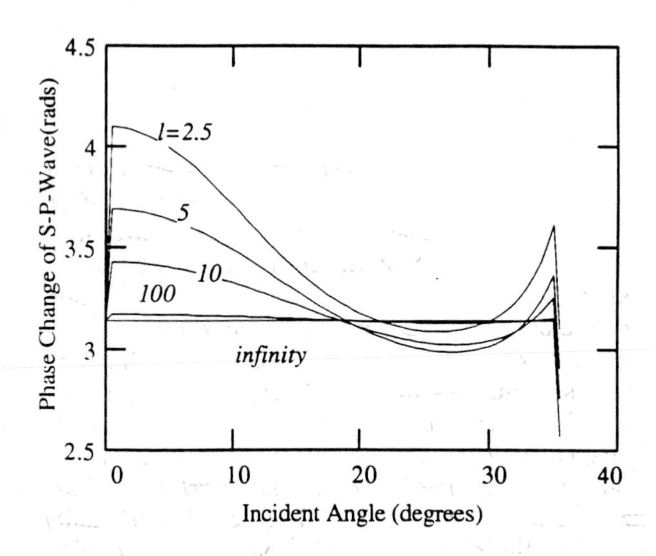

Fig. 6.12. Phase changes of the reflected P-wave with respect to the incident S-wave in the scale case of M=2 with l, the dimensionless scale factor as a parameter.

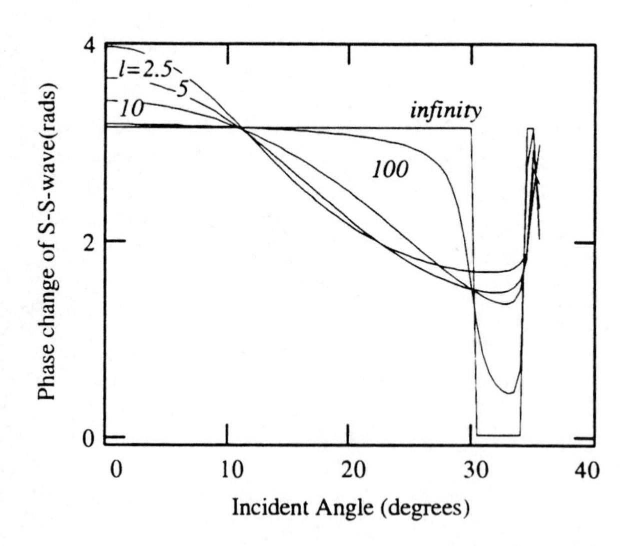

Fig. 6.13. Phase changes of the reflected S-wave with respect to the incident S-wave in the scale case of M=2 with l, the dimensionless scale factor as a parameter.

change of π around 30°–35°, at which the mode conversion of reflected S wave occurs.

6.6 Closure

In this chapter, doublet mechanics has been used to analyze the elastic wave reflection at discrete free-boundary interfaces. The amplitude ratios of reflected waves to incident waves and phase shifts of the reflected waves have been obtained analytically. It was found that the results of first order approximation, which corresponds to the situation of nonscale, isotropic solids, are consistent with those obtained using classical continuum theory.

Upon accounting for scale effects, novel results were established that have no counterpart in classical elastodynamics. In particular, it was found that the critical angle of mode conversion, the amplitude ratios, and the phase shifts in reflection phenomena are dependent on a dimensionless scale parameter. Such dependence is more pronounced at shorter dimensionless wavelengths, that is, for larger discrete sizes for a fixed wavelength. The non-dimensionality of the scale parameter renders the present results applicable to materials with significantly different typical microstructural dimensions, as long as the wavelengths are proportionately scaled. The classical elastodynamic reflection results were found to hold as an excellent approximation in the wavelength-to-discrete size ratio of order 100.

7. Surface Waves – Difference Equations

V. T. Granik

7.1 Introduction

The first study of surface waves dates back to the end of the nineteenth century with an article by Lord Rayleigh (1885). He considered a plane wave traveling along the free boundary of a solid half-space and decaying exponentially with depth. The continuum solid was assumed to be linear-elastic, homogeneous and isotropic. It was established that the velocity of such a wave is independent of the wavelength, i.e., the classical Rayleigh surface wave (RSW) is nondispersive. Because of propagating predominantly in a thin surface layer, the RSW can travel over long distances without considerable attenuation. Due to this, and some other unique properties, the RSW and other surface waves (SW) have found many uses in geophysics, seismology, signal processing and nondestructive evaluation. A review of these applications can be found in the edited volumes of Miklowitz and Achenbach (1978), Oliner (1978), Blakemore and Georgiou (1988), Mal and Singh (1991) Datta et al (1990) and Achenbach (1993).

The technical and physical applications imparted a powerful impetus to SW investigations. Since Lord Rayleigh's seminal work, this scientific field has been extended far beyond the classical half-infinite continuum, especially over the past several decades. Now it includes solids of different shapes, sizes, internal structures and physical properties. Basic aspects of the SW studies are covered in various monographs by Ewing et al (1957), Kolsky (1963), Viktorov (1967), Pilant (1979), Achenbach (1980), Brekhoskikh (1980), Hudson (1980), Miklowitz (1980), Ben-Menahem and Singh (1981), Malischewsky (1987) and others. The particular phenomena of the RSW propagation (diffraction, reflection, refraction, scattering, dispersion) as well as higher-order effects due to anisotropy, inhomogeneity, etc., are considered in numerous articles.

Until now almost all theoretical researches into the SW have been undertaken on the assumption of local elasticity. As has already been mentioned in Chap. 3, the local theories of elasticity disregard the underlying discrete microstructure of solids at all levels, from atoms in crystals to the grains of bulk materials and the blocks of the Earth's crust. Considering all solids as continua, the local models are capable of revealing a *macroscopic* dispersion that derives from finite macroscopic dimensions of both solids (rods, plates, shells, etc.) and their internal elements (layers, inclusions, cavities, cracks, and so on). Due to this point, macroscopic dispersion is encountered over the band of relatively long wavelengths which are comparable with the above macroscopic dimensions.

There exists, however, another kind of SW dispersion, a *microscopic* one. This kind of dispersion is induced by scale parameters of microstructure that are small in comparison with dimensions of the solid. These parameters may be small interatomic distances in crystals, or sizable grains in granular arrays, or gigantic blocks of the Earth's crust. A study of microscopic dispersion is impossible by the local theories of elasticity.

Here we address the problem of the RSW on the basis of doublet mechanics (DM). Being a nonlocal theory of elasticity, DM enables one to take into account the aforementioned scale parameters and, thus, open the way for studying microscopic dispersion. It should be noted, however, that this chapter is not an extended research on surface waves. It is to be regarded only as an introduction to the subject limited to consideration of a typical problem: the propagation of the RSW in an elastic half-space. But unlike the classical continuum, this half- space is assumed to have a certain (in particular, a simple cubic) microstructure.

The problem of propagating the RSW in media with cubic microstructure is not new *per se*. It first attracted attention about forty years ago when Stoneley (1955) calculated secular equations of the RSW traveling in an elastic cubic half-space assumed to be an anisotropic continuum. Within the framework of continuum mechanics, surface waves in media with cubic microstructure were then considered by Gazis et al (1960), Buchwald and Davis (1963), Rollins et al (1968), Chadwick and Smith (1982), Royer and Dieulesaint (1984) to mention a few. In terms of continuum mechanics such media prove to be nondispersive. Recently, a series of studies of the RSW dispersion in so-called superlattices have been performed drawing on the conventional (local) theory of elasticity. The superlattices consist of two different alternating thin layers, each layer usually being a material of isotropic or cubic symmetry. The number of layers is taken large enough so that the layer thickness may be deemed as a scale parameter of the superstructure resulting in an intermediate (between macro- and microscopic) dispersion of the RSW. Kueny and Grimsditch (1982) calculated the velocities of the RSW in a superlattice of up to 1000 alternating layers of Nb and Cu deposited on a substrate. Djafari-Rouhani et al (1984) addressed the propagation of the RSW on a superlattice cut normal to its laminations.

Some researches on surface waves in cubic crystals with central interatomic forces have been conducted on the basis of lattice dynamics (LD). Using an atomic model of LD, Dobrzynski et al (1984) studied surface-localized phonons in a three-dimensional superlattice built up from two different alternating simple cubic lattices. A number of surface wave problems were considered in the fundamental monograph on LD written by Maradudin et al (1971,

Chapter IX and references therein). Within the scope of a nonlocal theory of elasticity different from LD and DM, the problem of propagating the RSW in a homogeneous isotropic medium with central binary interaction was treated by Kunin (1983).

We now refer to the problem of the RSW in order to show the subtle capabilities of DM which are inaccessible to local continuum theories. We restrict ourselves to the lattice model with nearest-neighbor interactions. Unlike LD, the underlying particles of a solid have two independent degrees of freedom, translations and rotations. This makes the particles interact not only through central but also through shear microforces. To obtain a solution of the problem we apply a variant of DM: *difference* governing equations, in contrast to *differential* governing equations in the previous chapters. These doublet mechanical equations enable us to reveal the microscopic dispersion of the classical RSW over the band of relatively short wavelengths. In passing, microscopic dispersion of plane P and S waves is established. We also compare the results obtained with those yielded by the classical (local) theory of elasticity, as well as by another nonlocal model of elasticity (Kunin 1983).

7.2 Geometry, Kinematics and Microstrains

In what follows we apply the notions, terminology, notation, summation conventions and governing equations of DM as they are set forth in Chap. 1.

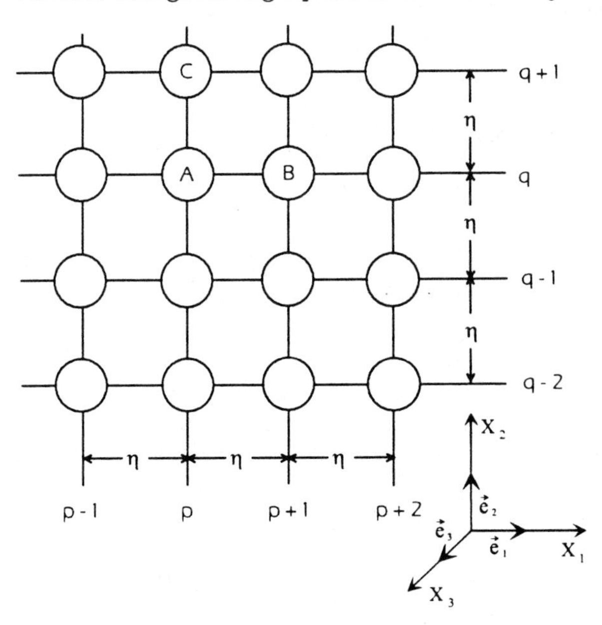

Fig. 7.1. Elastic, half-infinite solid with simple cubic microstructure

We consider an elastic half-infinite solid with a simple cubic lattice with a parameter η (Fig. 7.1). The initial and current configurations of the solid are identified in the rectangular Cartesian frame of reference $\{x_i\}$ with a right-handed unit vector basis $\{\mathbf{e}_i\}$ obeying the relations

$$\mathbf{e}_i \cdot \mathbf{e}_j = \delta_{ij}, \tag{7.1}$$

$$\mathbf{e}_i = \epsilon_{ijk}\, \mathbf{e}_j \times \mathbf{e}_k. \tag{7.2}$$

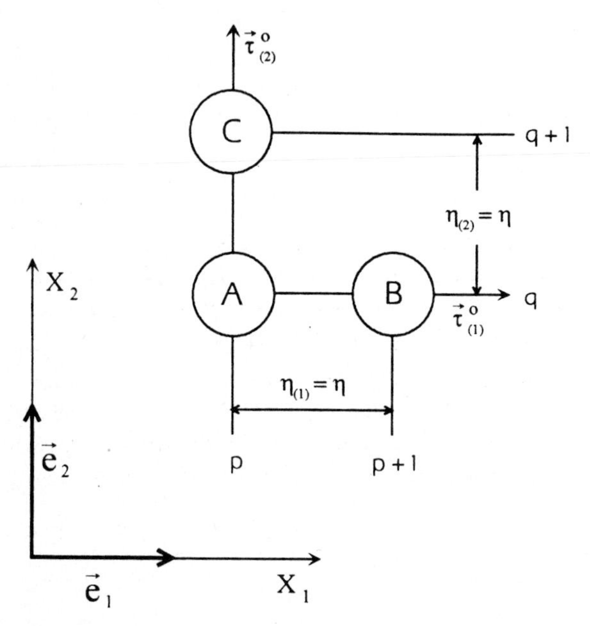

Fig. 7.2. Underlying microstructural element (bundle of the solid consisting of two doublets (A, B) and (A, C))

We consider a plane problem with a plane two-valence ($n = 2$) bundle T_n of doublets (A, B_α), $\alpha = 1, 2$, or simply (A, B) and (A, C) (Fig. 7.2). The unit vector-directors $\boldsymbol{\tau}^\circ_\alpha$ of the bundle T_n are, in general, determined as

$$\boldsymbol{\tau}^\circ_\alpha = \tau^\circ_{\alpha i}\, \mathbf{e}_i. \tag{7.3}$$

For the cubic microstructure in question, the doublets (A, B) and (A, C) are parallel to the unit vectors $\mathbf{e}_1$ and $\mathbf{e}_2$, respectively (Fig. 7.2). Therefore in eqn. (7.3) $\tau^\circ_{\alpha i} = \delta_{\alpha i}$, which results in the relations

$$\boldsymbol{\tau}^\circ_{(1)} = \mathbf{e}_1, \qquad \boldsymbol{\tau}^\circ_{(2)} = \mathbf{e}_2, \qquad \boldsymbol{\tau}^\circ_{(3)} = \mathbf{0}. \tag{7.4}$$

Hereafter the numerical values of the Greek indices are placed between parentheses to distinguish them from the Latin indices.

The kinematics of an arbitrary particle A of the solid is determined by the independent translations $\mathbf{u}$ and rotations $\boldsymbol{\phi}$. In general,

$$\mathbf{u} = u_i\,\mathbf{e}_i, \qquad \boldsymbol{\phi} = \phi_i\,\mathbf{e}_i. \tag{7.5}$$

In the plane problem, $u_3 = \phi_1 = \phi_2 = 0$, and the relations (7.5) reduce to

$$\mathbf{u} = u_1\,\mathbf{e}_1 + u_2\,\mathbf{e}_2, \tag{7.6}$$

$$\boldsymbol{\phi} = \varphi\,\mathbf{e}_3 \tag{7.7}$$

where $\varphi \equiv \phi_3$. When passing from particle A to the adjacent particle $B_\alpha \in (A, B_\alpha)$, the vectors $\mathbf{u}$ and $\boldsymbol{\phi}$ gain increments $\Delta\mathbf{u}_\alpha$ and $\Delta\boldsymbol{\phi}_\alpha$, respectively. The increments $\Delta\mathbf{u}_\alpha$ and $\Delta\boldsymbol{\phi}_\alpha$ can be represented as

$$\Delta\mathbf{u}_\alpha = \Delta u_{\alpha i}\,\mathbf{e}_i, \qquad \Delta\boldsymbol{\phi}_\alpha = \Delta\phi_{\alpha i}\,\mathbf{e}_i. \tag{7.8}$$

In terms of eqns. (7.6) and (7.7), eqns. (7.8) take the form

$$\Delta\mathbf{u}_\alpha = \Delta u_{\alpha 1}\,\mathbf{e}_1 + \Delta u_{\alpha 2}\,\mathbf{e}_2, \tag{7.9}$$

$$\Delta\boldsymbol{\phi}_\alpha = \Delta\varphi_\alpha\,\mathbf{e}_3. \tag{7.10}$$

Now we consider the doublet microstrains of elongation $\boldsymbol{\epsilon}_\alpha$, torsion $\boldsymbol{\mu}_\alpha$ and shear $\boldsymbol{\gamma}_\alpha$. Given that the scale parameters $\eta_\alpha = \eta,\ \alpha = 1, 2$, the magnitude ϵ_α of the axial microstrain $\boldsymbol{\epsilon}_\alpha = \epsilon_\alpha\,\boldsymbol{\tau}_\alpha^\circ$ is expressed by eqn. (1.11)

$$\epsilon_\alpha = \eta^{-1}\,(\Delta\mathbf{u}_\alpha \cdot \boldsymbol{\tau}_\alpha^\circ). \tag{7.11}$$

In view of the relations (7.1), (7.4) and (7.9), eqns. (7.11) become

$$\epsilon_{(1)} = \eta^{-1}\Delta u_{(1)1}, \qquad \epsilon_{(2)} = \eta^{-1}\Delta u_{(2)2}. \tag{7.12}$$

The magnitude μ_α of the torsion microstrain $\boldsymbol{\mu}_\alpha = \mu_\alpha\,\boldsymbol{\tau}_\alpha^\circ$ is determined by (eqn. (1.16))

$$\mu_\alpha = \eta^{-1}\,(\Delta\boldsymbol{\phi}_\alpha \cdot \boldsymbol{\tau}_\alpha^\circ). \tag{7.13}$$

If we take account of the relations (7.1), (7.4) and (7.10), then eqn. (7.13) reduces to the zero identity $\mu_\alpha \equiv 0$. So, in the plane problem, there are no microstrains of torsion μ_α.

The shear microstrain $\boldsymbol{\gamma}_\alpha$ is given by (eqn. (1.21))

$$\boldsymbol{\gamma}_\alpha = \left[\boldsymbol{\psi}_\alpha - \left(\boldsymbol{\phi} + \frac{1}{2}\Delta\boldsymbol{\phi}_\alpha\right)\right] \times \boldsymbol{\tau}_\alpha^\circ. \tag{7.14}$$

In eqn. (7.14), $\boldsymbol{\psi}_\alpha$ is a small angle of the initial vector-director $\boldsymbol{\tau}_\alpha^\circ$ with its actual position $\boldsymbol{\tau}_\alpha$. Since the axial microstrains ϵ_α are assumed to be small ($|\epsilon_\alpha| \ll 1$) and the scale parameters $\eta_\alpha = \eta$, eqn. (1.9) takes the form

$$\boldsymbol{\tau}_\alpha = \boldsymbol{\tau}_\alpha^\circ + \eta^{-1}\,\Delta\mathbf{u}_\alpha. \tag{7.15}$$

With regard to the relations (7.7), (7.10) and (1.22), eqn. (7.14) takes the form

$$\boldsymbol{\gamma}_\alpha = \left[\boldsymbol{\tau}_\alpha^\circ \times \boldsymbol{\tau}_\alpha - \left(\varphi + \frac{1}{2}\Delta\varphi_\alpha\right)\mathbf{e}_3\right] \times \boldsymbol{\tau}_\alpha^\circ. \tag{7.16}$$

In eqn. (7.16) the double cross product $(\boldsymbol{\tau}_\alpha^\circ \times \boldsymbol{\tau}_\alpha) \times \boldsymbol{\tau}_\alpha^\circ$ can be written as

$$(\boldsymbol{\tau}_\alpha^\circ \times \boldsymbol{\tau}_\alpha) \times \boldsymbol{\tau}_\alpha^\circ = (\boldsymbol{\tau}_\alpha^\circ \cdot \boldsymbol{\tau}_\alpha^\circ)\, \boldsymbol{\tau}_\alpha - (\boldsymbol{\tau}_\alpha^\circ \cdot \boldsymbol{\tau}_\alpha)\, \boldsymbol{\tau}_\alpha^\circ \equiv \boldsymbol{\tau}_\alpha - (\boldsymbol{\tau}_\alpha^\circ \cdot \boldsymbol{\tau}_\alpha)\, \boldsymbol{\tau}_\alpha^\circ . \tag{7.17}$$

Here the dot product $\boldsymbol{\tau}_\alpha^\circ \cdot \boldsymbol{\tau}_\alpha^\circ = 1$ because $|\boldsymbol{\tau}_\alpha^\circ| = 1$. In view of eqns. (7.11) and (7.15), eqn. (7.17) becomes

$$(\boldsymbol{\tau}_\alpha^\circ \times \boldsymbol{\tau}_\alpha) \times \boldsymbol{\tau}_\alpha^\circ = \eta^{-1}\, \Delta \mathbf{u}_\alpha - \epsilon_\alpha\, \boldsymbol{\tau}_\alpha^\circ . \tag{7.18}$$

Substituting $(\boldsymbol{\tau}_\alpha^\circ \times \boldsymbol{\tau}_\alpha) \times \boldsymbol{\tau}_\alpha^\circ$ from eqn. (7.18) into eqn. (7.16), we have

$$\boldsymbol{\gamma}_\alpha = \eta^{-1}\, \Delta \mathbf{u}_\alpha - \epsilon_\alpha\, \boldsymbol{\tau}_\alpha^\circ - \left(\varphi + \frac{1}{2}\, \Delta\varphi_\alpha\right) \mathbf{e}_3 \times \boldsymbol{\tau}_\alpha^\circ . \tag{7.19}$$

Equations (7.2) and (7.4) bring about the following expressions for the last term in eqn. (7.19):

$$\left(\varphi + \frac{1}{2}\, \Delta\varphi_{(1)}\right) \mathbf{e}_3 \times \boldsymbol{\tau}_{(1)}^\circ = \left(\varphi + \frac{1}{2}\, \Delta\varphi_{(1)}\right) \mathbf{e}_2 , \tag{7.20}$$

$$\left(\varphi + \frac{1}{2}\, \Delta\varphi_{(2)}\right) \mathbf{e}_3 \times \boldsymbol{\tau}_{(2)}^\circ = -\left(\varphi + \frac{1}{2}\, \Delta\varphi_{(2)}\right) \mathbf{e}_1 . \tag{7.21}$$

From eqn. (7.14) it follows that

$$\boldsymbol{\gamma}_\alpha \cdot \boldsymbol{\tau}_\alpha^\circ = \left(\left[\boldsymbol{\psi}_\alpha - \left(\boldsymbol{\phi} + \frac{1}{2}\, \Delta\boldsymbol{\phi}_\alpha\right)\right] \times \boldsymbol{\tau}_\alpha^\circ\right) \cdot \boldsymbol{\tau}_\alpha^\circ \equiv 0 . \tag{7.22}$$

This identity means that the microshear $\boldsymbol{\gamma}_\alpha$ of any doublet (A, B_α) is perpendicular to the unit vector-director $\boldsymbol{\tau}_\alpha^\circ$ of this doublet: $\boldsymbol{\gamma}_\alpha \perp \boldsymbol{\tau}_\alpha^\circ$. Along with eqn. (7.4), the identity (7.22) also means that $\boldsymbol{\gamma}_{(1)} \perp \mathbf{e}_1$ and $\boldsymbol{\gamma}_{(2)} \perp \mathbf{e}_2$. Hence the general representation of the vectors $\boldsymbol{\gamma}_\alpha$

$$\boldsymbol{\gamma}_\alpha = \gamma_{\alpha i}\, \mathbf{e}_i \tag{7.23}$$

assumes the form

$$\boldsymbol{\gamma}_{(1)} = \gamma_{(1)2}\, \mathbf{e}_2 , \qquad \boldsymbol{\gamma}_{(2)} = \gamma_{(2)1}\, \mathbf{e}_1 . \tag{7.24}$$

Now the relations (7.9), (7.12), (7.20), (7.21) and (7.24) allow us to write eqn. (7.19) as follows:

$$\gamma_{(1)2} = \eta^{-1}\, \Delta u_{(1)2} - \left(\varphi + \frac{1}{2}\, \Delta\varphi_{(1)}\right) , \tag{7.25}$$

$$\gamma_{(2)1} = \eta^{-1}\, \Delta u_{(2)1} - \left(\varphi + \frac{1}{2}\, \Delta\varphi_{(2)}\right) . \tag{7.26}$$

7.3 Microstresses and Constitutive Equations

Any deformation of a solid which induces the microstrains ϵ_α, μ_α and $\gamma_{\alpha i}$, simultaneously generates corresponding microstresses p_α, m_α and $t_{\alpha i}$. The relationship between the microstrains and the microstresses is given by the constitutive equations (Chap. 2)

$$p_\alpha = \sum_{\beta=1}^{n} A_{\alpha\beta}\, \epsilon_\beta + J_\alpha\, \Theta, \tag{7.27}$$

$$m_\alpha = \sum_{\beta=1}^{n} E_{\alpha\beta}\, \mu_\beta, \tag{7.28}$$

$$t_{\alpha i} = \sum_{\beta=1}^{n} I_{\alpha\beta ij}\, \gamma_{\beta j}. \tag{7.29}$$

We will apply these equations in a simplified form by taking the following assumptions:

1. Thermal effects are neglected ($J_\alpha\, \Theta = 0$).
2. Particle interactions are local ($A_{\alpha\beta} = A_\alpha\, \delta_{\alpha\beta}$, $E_{\alpha\beta} = E_\alpha\, \delta_{\alpha\beta}$, $I_{\alpha\beta ij} = I_\alpha\, \delta_{\alpha\beta}\, \delta_{ij}$).

   With these two assumptions, eqns. (7.27)–(7.29) become

$$p_\alpha = A_\alpha\, \epsilon_\alpha, \qquad m_\alpha = E_\alpha\, \mu_\alpha, \qquad t_{\alpha i} = I_\alpha\, \gamma_{\alpha i}. \tag{7.30}$$

3. Particle interactions are homogeneous (the physical constants $A_\alpha = A_\circ$, $E_\alpha = E_\circ$, $I_\alpha = I_\circ$ for any $\alpha = 1, 2$). Therefore, the elastic deformation of the solid is defined by the three micromoduli of elasticity $A_\circ$, $E_\circ$ and $I_\circ$. In view of this assumption, eqns. (7.30) take the form

$$p_\alpha = A_\circ\, \epsilon_\alpha, \qquad m_\alpha = E_\circ\, \mu_\alpha, \qquad t_{\alpha i} = I_\circ\, \gamma_{\alpha i}. \tag{7.31}$$

7.4 Difference Kinematic and Constitutive Equations

Consider an arbitrary bundle of the cubic microstructure in Fig. 7.2 and the translations and rotations of the constituent particles A, B, C in Fig. 7.3. For the doublet (A, B), whose index $\alpha = 1$, we have the following increments of translations and rotations (Fig. 7.3):

$$\left\{ \begin{array}{c} \Delta u_{(1)1} \\ \Delta u_{(1)2} \\ \Delta \varphi_{(1)} \end{array} \right\} = \left\{ \begin{array}{c} U_B - U_A \\ V_B - V_A \\ \varphi_B - \varphi_A \end{array} \right\} \tag{7.32}$$

For the doublet (A, C), whose index $\alpha = 2$, we obtain

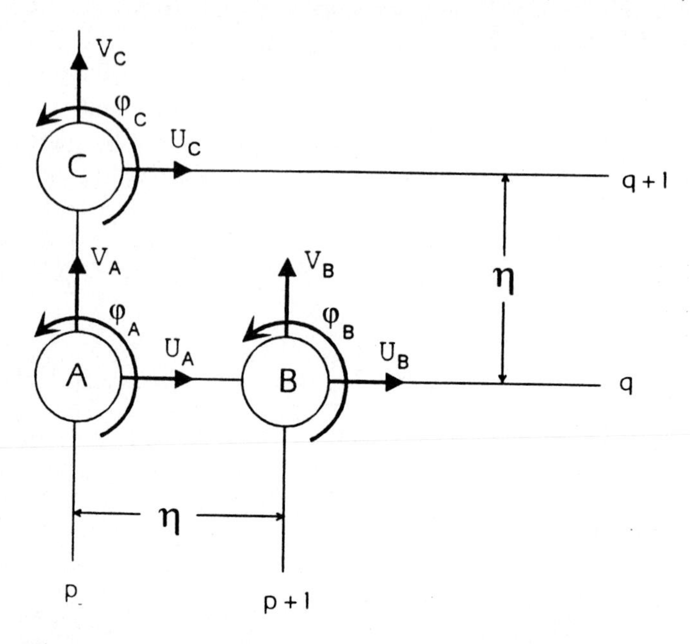

Fig. 7.3. Translations and rotations of the constituent particles of the underlying bundle

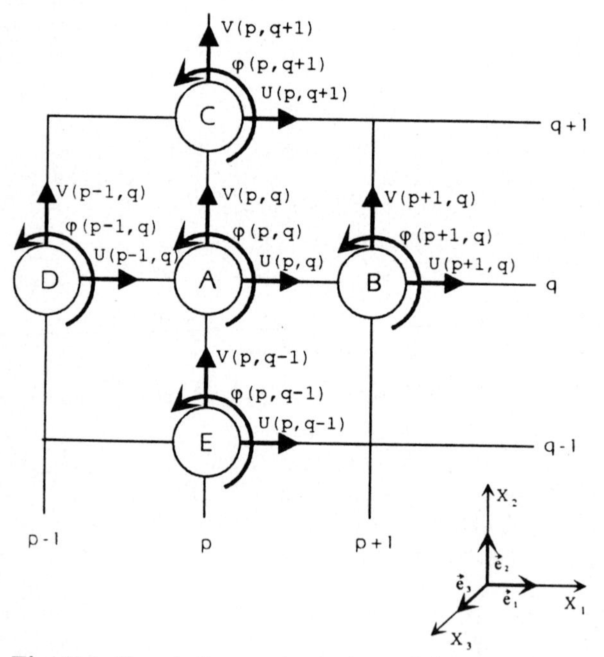

Fig. 7.4. Translations and rotations of the particles adjacent to an arbitrary particle *A* of the solid

$$\left\{ \begin{array}{c} \Delta u_{(2)1} \\ \Delta u_{(2)2} \\ \Delta \varphi_{(2)} \end{array} \right\} = \left\{ \begin{array}{c} U_C - U_A \\ V_C - V_A \\ \varphi_C - \varphi_A \end{array} \right\} \tag{7.33}$$

Together with eqns. (7.12), (7.25) and (7.26), the relations (7.32) and (7.33) determine the microstrains of elongation and shear of the doublets (A, B) and (A, C):

$$\left\{ \begin{array}{c} \epsilon_{(1)} \\ \epsilon_{(2)} \end{array} \right\} \equiv \left\{ \begin{array}{c} \epsilon_{AB} \\ \epsilon_{AC} \end{array} \right\} = \frac{1}{\eta} \left\{ \begin{array}{c} U_B - U_A \\ U_C - U_A \end{array} \right\} \tag{7.34}$$

$$\left\{ \begin{array}{c} \gamma_{(1)2} \\ \gamma_{(2)1} \end{array} \right\} \equiv \left\{ \begin{array}{c} \gamma_{AB} \\ \gamma_{AC} \end{array} \right\} = \frac{1}{\eta} \left\{ \begin{array}{c} V_B - V_A - (\bar{\varphi}_A + \bar{\varphi}_B) \\ U_C - U_A + (\bar{\varphi}_A + \bar{\varphi}_C) \end{array} \right\} \tag{7.35}$$

where $\bar{\varphi} \equiv \eta\varphi/2$. Now we consider Fig. 7.4 which shows an arbitrary particle A with the doublets that belong to it: (A, B), (A, C), (A, D), (A, E). In terms of eqns. (7.34) and (7.35) and the discrete coordinates (p, q) (Fig. 7.4) we obtain the microstrains of elongation and shear of all the adjacent doublets:

$$\{E\} \equiv \left\{ \begin{array}{c} \epsilon_{AB} \\ \epsilon_{AC} \\ \epsilon_{AD} \\ \epsilon_{AE} \end{array} \right\} = \frac{1}{\eta} \left\{ \begin{array}{c} U(p+1,q) - U(p,q) \\ V(p,q+1) - V(p,q) \\ U(p,q) - U(p-1,q) \\ V(p,q) - V(p,q-1) \end{array} \right\} \tag{7.36}$$

$$\begin{aligned} \{\Gamma\} &\equiv \left\{ \begin{array}{c} \gamma_{AB} \\ \gamma_{AC} \\ \gamma_{AD} \\ \gamma_{AE} \end{array} \right\} \\ &= \frac{1}{\eta} \left\{ \begin{array}{c} V(p+1,q) - V(p,q) - \bar{\varphi}(p,q) - \bar{\varphi}(p+1,q) \\ U((p,q+1) - U(p,q) + \bar{\varphi}(p,q) + \bar{\varphi}(p,q+1) \\ V(p,q) - V(p-1,q) - \bar{\varphi}(p,q) - \bar{\varphi}(p-1,q) \\ U(p,q) - U(p,q-1) + \bar{\varphi}(p,q) + \bar{\varphi}(p,q-1) \end{array} \right\} \end{aligned} \tag{7.37}$$

Equations (7.31) and the relations (7.36) and (7.37) result in the constitutive equations for the microstresses in the doublets (A, B), (A, C), (A, D), (A, E): $\{P\} = A_\circ \{E\}$ and $\{T\} = I_\circ \{\Gamma\}$, or

$$\{P\} \equiv \left\{ \begin{array}{c} p_{AB} \\ p_{AC} \\ p_{AD} \\ p_{AE} \end{array} \right\} = \frac{A_\circ}{\eta} \left\{ \begin{array}{c} U(p+1,q) - U(p,q) \\ V(p,q+1) - V(p,q) \\ U(p,q) - U(p-1,q) \\ V(p,q) - V(p,q-1) \end{array} \right\} \tag{7.38}$$

$$\{T\} \equiv \begin{Bmatrix} t_{AB} \\ t_{AC} \\ t_{AD} \\ t_{AE} \end{Bmatrix}$$

$$= \frac{I_\circ}{\eta} \begin{Bmatrix} V(p+1,q) - V(p,q) - \bar{\varphi}(p,q) - \bar{\varphi}(p+1,q) \\ U((p,q+1) - U(p,q) + \bar{\varphi}(p,q) + \bar{\varphi}(p,q+1) \\ V(p,q) - V(p-1,q) - \bar{\varphi}(p,q) - \bar{\varphi}(p-1,q) \\ U(p,q) - U(p,q-1) + \bar{\varphi}(p,q) + \bar{\varphi}(p,q-1) \end{Bmatrix} \tag{7.39}$$

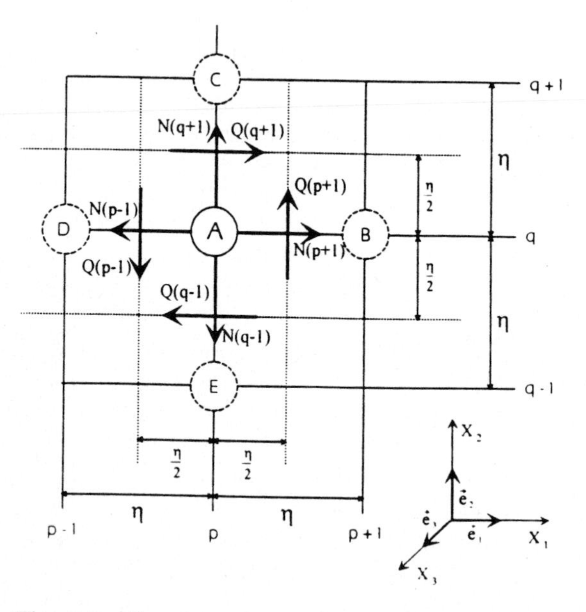

Fig. 7.5. Microforces acting upon an arbitrary particle A of the solid

The microstresses $\{P\}$ and $\{T\}$ correspond to the internal axial and shear microforces $\{N\} = \eta^2 \{P\}$ and $\{Q\} = \eta^2 \{T\}$ respectively, η^2 being the area of the face of a simple cubic cell. We have (Fig. 7.5)

$$\{N\} \equiv \begin{Bmatrix} N_{AB} \\ N_{AC} \\ N_{AD} \\ N_{AE} \end{Bmatrix} \equiv \begin{Bmatrix} N(p+1) \\ N(q+1) \\ N(p-1) \\ N(q-1) \end{Bmatrix}$$

$$= A_\circ \eta \begin{Bmatrix} U(p+1,q) - U(p,q) \\ V(p,q+1) - V(p,q) \\ U(p,q) - U(p-1,q) \\ V(p,q) - V(p,q-1) \end{Bmatrix} \tag{7.40}$$

$$\{Q\} \equiv \left\{ \begin{array}{c} Q_{AB} \\ Q_{AC} \\ Q_{AD} \\ Q_{AE} \end{array} \right\} \equiv \left\{ \begin{array}{c} Q(p+1) \\ Q(q+1) \\ Q(p-1) \\ Q(q-1) \end{array} \right\}$$

$$= I_\circ \eta \left\{ \begin{array}{c} V(p+1,q) - V(p,q) - \bar{\varphi}(p,q) - \bar{\varphi}(p+1,q) \\ U((p,q+1) - U(p,q) + \bar{\varphi}(p,q) + \bar{\varphi}(p,q+1) \\ V(p,q) - V(p-1,q) - \bar{\varphi}(p,q) - \bar{\varphi}(p-1,q) \\ U(p,q) - U(p,q-1) + \bar{\varphi}(p,q) + \bar{\varphi}(p,q-1) \end{array} \right\} \quad (7.41)$$

7.5 Difference Equations of Motion

Let us return to an arbitrary particle A whose position in the initial configuration is identified by the lattice coordinates (p,q). The equations of motion of the particle can be written as

$$\mathbf{F}(p,q;t) = m\,\frac{d^2\mathbf{u}(p,q;t)}{dt^2}, \quad (7.42)$$

$$\mathbf{M}(p,q;t) = J\,\frac{d^2\boldsymbol{\varphi}(p,q;t)}{dt^2} \equiv J\,\frac{2}{\eta}\,\frac{d^2\bar{\boldsymbol{\varphi}}(p,q;t)}{dt^2} \quad (7.43)$$

where t is time, m and J are the mass and the axial moment of inertia of the particle, $\mathbf{F}(p,q;t)$ and $\mathbf{M}(p,q;t)$ are the total microforce and the total micromoment of microforces acting on the particle. In the notations $\mathbf{F}(p,q;t)$, $\mathbf{M}(p,q;t)$, $\mathbf{u}(p,q;t)$, $\boldsymbol{\varphi}\,(p,q;t)$, $\bar{\boldsymbol{\varphi}}(p,q;t)$, the symbol t is separated from p, q in order to emphasize an essential difference between them. The point is that the functions $\mathbf{F}(p,q;t)$, $\mathbf{M}(p,q;t)$, $\mathbf{u}(p,q;t)$, $\boldsymbol{\varphi}(p,q;t)$, $\bar{\boldsymbol{\varphi}}(p,q;t)$ are *continuous* in t and *discrete* in p, q. Hereafter, for the sake of simplicity, the argument t will generally be omitted, although implied.

In the plane problem the vectors $\mathbf{u}(p,q)$, $\boldsymbol{\varphi}(p,q)$, $\mathbf{F}(p,q)$, $\mathbf{M}(p,q)$ are (Figs. 7.4 and 7.5)

$$\mathbf{u}(p,q) = U(p,q)\,\mathbf{e}_1 + V(p,q)\,\mathbf{e}_2, \quad (7.44)$$

$$\boldsymbol{\varphi}(p,q) = \varphi(p,q)\,\mathbf{e}_3, \quad (7.45)$$

$$\begin{aligned} \mathbf{F}(p,q) &= [N(p+1) - N(p-1) + Q(q+1) - Q(q-1)]\,\mathbf{e}_1 \\ &\quad + [N(q+1) - N(q-1) + Q(p+1) - Q(p-1)]\,\mathbf{e}_2, \end{aligned} \quad (7.46)$$

$$\mathbf{M}(p,q) = \frac{1}{2}\,\eta[Q(p+1) + Q(p-1) - Q(q+1) - Q(q-1)]\,\mathbf{e}_3. \quad (7.47)$$

Equations (7.46) and (7.47) are valid for any particle A inside the solid when $q = 1, 2, \ldots$ or $x_2 = \eta, 2\eta, \ldots$ (Fig. 7.1). If the particle is located on the surface $q = 0$ and $x_2 = 0$, the internal microforces $N(q-1)$, $Q(q-1)$ are absent and eqns. (7.46) and (7.47) become

$$\begin{aligned}\mathbf{F}(p,q) &= [N(p+1) - N(p-1) + Q(q+1)]\,\mathbf{e}_1 \\ &\quad + [N(q+1) + Q(p+1) - Q(p-1)]\,\mathbf{e}_2,\end{aligned} \tag{7.48}$$

$$\mathbf{M}(p,q) = \frac{1}{2}\,\eta\,[Q(p+1) + Q(p-1) - Q(q+1)]\,\mathbf{e}_3. \tag{7.49}$$

It should be noted that the vectors $\mathbf{F}(p,q)$ and $\mathbf{M}(p,q)$ have not included external (body) microforces because they usually are unessential in the wave studies.

We assume that the constituent particles of the solid are spherical, η being their diameter. The axial moment of inertia J of such a particle is

$$J = \frac{m\eta^2}{10}. \tag{7.50}$$

We take the mass

$$m = \eta^3\,\rho \tag{7.51}$$

where η^3 is the volume of a cubic cell, ρ is the average mass density of the solid. With regard to eqn. (7.51), eqn. (7.50) assumes the form

$$J = \frac{\eta^5\,\rho}{10}. \tag{7.52}$$

7.6 Difference Governing Equations

The above relations suffice to obtain the governing equations of the solid in terms of the particle translations and rotations. For this purpose, we substitute into the equations of motion (7.42) and (7.43) the displacements $\mathbf{u}(p,q)$ and rotations $\varphi(p,q)$ from eqns. (7.44) and (7.45), and the dynamic vectors $\mathbf{F}(p,q)$ and $\mathbf{M}(p,q)$ from eqns. (7.46)–(7.49). Then, in view of the constitutive relations (7.40) and (7.41), we obtain the difference governing equations:

1) For the particles inside the solid ($q = 1, 2, \ldots,\ x_2 = \eta q,\ 2\eta q, \ldots$)

$$\begin{aligned}&U(p+1,q) - 2(1+\vartheta)\,U(p,q) + U(p-1,q) \\ &\quad +\vartheta\,[U(p,q+1) + U(p,q-1) + \bar{\varphi}(p,q+1) - \bar{\varphi}(p,q-1)] \\ &\quad = \left(\frac{\eta}{C}\right)^2 \frac{d^2U(p,q)}{dt^2},\end{aligned} \tag{7.53}$$

$$\begin{aligned}&V(p,q+1) - 2(1+\vartheta)\,V(p,q) + V(p,q-1) \\ &\quad +\vartheta\,[V(p+1,q) + V(p-1,q) - \bar{\varphi}(p+1,q) + \bar{\varphi}(p-1,q)] \\ &\quad = \left(\frac{\eta}{C}\right)^2 \frac{d^2V(p,q)}{dt^2},\end{aligned} \tag{7.54}$$

$$\begin{aligned}&U(p,q+1) - U(p,q-1) - V(p+1,q) + V(p-1,q) + 4\bar{\varphi}(p,q) \\ &\quad +\bar{\varphi}(p+1,q) + \bar{\varphi}(p-1,q) + \bar{\varphi}(p,q+1) + \bar{\varphi}(p,q-1) \\ &\quad = -0.4\,\vartheta^{-1}\left(\frac{\eta}{C}\right)^2 \frac{d^2\bar{\varphi}(p,q)}{dt^2}.\end{aligned} \tag{7.55}$$

2) For the particles on the plane boundary of the solid ($q = 0$, $x_2 = 0$)

$$U(p+1,q) - (2+\vartheta)\,U(p,q) + U(p-1,q) \\ + \vartheta\,[U(p,q+1) + \bar{\varphi}(p,q) + \bar{\varphi}(p,q+1)] \\ = \left(\frac{\eta}{C}\right)^2 \frac{d^2U(p,q)}{dt^2}, \tag{7.56}$$

$$V(p,q+1) - (1+2\vartheta)\,V(p,q) + \vartheta\,[V(p+1,q) + V(p-1,q) \\ -\bar{\varphi}(p+1,q) + \bar{\varphi}(p-1,q)] \\ = \left(\frac{\eta}{C}\right)^2 \frac{d^2V(p,q)}{dt^2}, \tag{7.57}$$

$$U(p,q+1) - U(p,q) - V(p+1,q) + V(p-1,q) + 3\bar{\varphi}(p,q) \\ +\bar{\varphi}(p+1,q) + \bar{\varphi}(p-1,q) + \bar{\varphi}(p,q+1) \\ = -0.4\,\vartheta^{-1}\left(\frac{\eta}{C}\right)^2 \frac{d^2\bar{\varphi}(p,q)}{dt^2}. \tag{7.58}$$

In the above equations,

$$\vartheta = \frac{I_\circ}{A_\circ}, \qquad C = \sqrt{\frac{A_\circ}{\rho}}, \tag{7.59}$$

where C is a velocity whose physical meaning will be discussed below.

7.7 Continuum Approximation

The governing equations (7.53)–(7.58) are represented in terms of the discrete lattice coordinates (p,q). With respect to these coordinates the Cartesian coordinates (x_1, x_2) are $x_1 = p\eta$, $x_2 = q\eta$ (Fig. 7.1). Hence any lattice functions $f(p,q)$—the translations $U(p,q)$, $V(p,q)$, the rotations $\bar{\varphi}(p,q)$, and the other—may be mapped into $f(x_1,x_2)$: $f(p,q) \Rightarrow f(x_1,x_2)$. Analogously, $f(p+1,q) \mapsto f(x_1+\eta,x_2)$, $f(p,q+1) \mapsto f(x_1,x_2+\eta)$, etc. Beyond the lattice points $x_1 = p\eta$, $x_2 = q\eta$, the functions $f(x_1,x_2)$ are indeterminate *per se* and might therefore be redefined. In DM the functions $f(x_1,x_2)$ are supposed to be continuous and repeatedly differentiable in all the region occupied by the solid under consideration. Nonetheless, in this chapter such an assumption is not necessary and we will take it for a while in order to compare the above discrete governing equations with those of the classical elastic continuum. The foregoing assumption permits expanding the particle translations and rotations into finite power series

$$\left\{\begin{array}{l} U(p+1,q;t) \\ V(p+1,q;t) \\ \bar{\varphi}(p+1,q;t) \end{array}\right\} \Rightarrow \left\{\begin{array}{l} U(x_1+\eta,x_2,t) \\ V(x_1+\eta,x_2,t) \\ \bar{\varphi}(x_1+\eta,x_2,t) \end{array}\right\}$$

$$= \left(1+\mathcal{L}_{11}^{(\kappa)}\right) \left\{\begin{array}{l} U(x_1,x_2,t) \\ V(x_1,x_2,t) \\ \bar{\varphi}(x_1,x_2,t) \end{array}\right\} \tag{7.60}$$

$$\left\{\begin{array}{l} U(p-1,q;t) \\ V(p-1,q;t) \\ \bar{\varphi}(p-1,q;t) \end{array}\right\} \Rightarrow \left\{\begin{array}{l} U(x_1-\eta,x_2,t) \\ V(x_1-\eta,x_2,t) \\ \bar{\varphi}(x_1-\eta,x_2,t) \end{array}\right\}$$

$$= \left(1+\mathcal{L}_{12}^{(\kappa)}\right) \left\{\begin{array}{l} U(x_1,x_2,t) \\ V(x_1,x_2,t) \\ \bar{\varphi}(x_1,x_2,t) \end{array}\right\} \tag{7.61}$$

$$\left\{\begin{array}{l} U(p,q+1;t) \\ V(p,q+1;t) \\ \bar{\varphi}(p,q+1;t) \end{array}\right\} \Rightarrow \left\{\begin{array}{l} U(x_1,x_2+\eta,t) \\ V(x_1,x_2+\eta,t) \\ \bar{\varphi}(x_1,x_2+\eta,t) \end{array}\right\}$$

$$= \left(1+\mathcal{L}_{21}^{(\kappa)}\right) \left\{\begin{array}{l} U(x_1,x_2,t) \\ V(x_1,x_2,t) \\ \bar{\varphi}(x_1,x_2,t) \end{array}\right\} \tag{7.62}$$

$$\left\{\begin{array}{l} U(p,q-1;t) \\ V(p,q-1;t) \\ \bar{\varphi}(p,q-1;t) \end{array}\right\} \Rightarrow \left\{\begin{array}{l} U(x_1,x_2-\eta,t) \\ V(x_1,x_2-\eta,t) \\ \bar{\varphi}(x_1,x_2-\eta,t) \end{array}\right\}$$

$$= \left(1+\mathcal{L}_{22}^{(\kappa)}\right) \left\{\begin{array}{l} U(x_1,x_2,t) \\ V(x_1,x_2,t) \\ \bar{\varphi}(x_1,x_2,t) \end{array}\right\} \tag{7.63}$$

where the differential operators $\mathcal{L}_{11}^{(\kappa)}$, $\mathcal{L}_{12}^{(\kappa)}$, $\mathcal{L}_{21}^{(\kappa)}$, $\mathcal{L}_{22}^{(\kappa)}$ $(M \geq 2)$ are defined by

$$\mathcal{L}_{11}^{(\kappa)} = \sum_{\kappa=1}^{M} \frac{\eta^\kappa}{\kappa!} \frac{\partial^\kappa}{\partial x_1^\kappa} \tag{7.64}$$

$$\mathcal{L}_{12}^{(\kappa)} = \sum_{\kappa=1}^{M} (-1)^\kappa \frac{\eta^\kappa}{\kappa!} \frac{\partial^\kappa}{\partial x_1^\kappa} \tag{7.65}$$

$$\mathcal{L}_{21}^{(\kappa)} = \sum_{\kappa=1}^{M} \frac{\eta^\kappa}{\kappa!} \frac{\partial^\kappa}{\partial x_2^\kappa} \tag{7.66}$$

$$\mathcal{L}_{22}^{(\kappa)} = \sum_{\kappa=1}^{M} (-1)^\kappa \frac{\eta^\kappa}{\kappa!} \frac{\partial^\kappa}{\partial x_2^\kappa}. \tag{7.67}$$

From eqns. (7.64)–(7.67) it follows that

$$\mathcal{L}_{11}^{(\kappa)} + \mathcal{L}_{12}^{(\kappa)} = 2\,\mathcal{L}_{11}^{(2\kappa)}, \tag{7.68}$$

$$\mathcal{L}_{11}^{(\kappa)} - \mathcal{L}_{12}^{(\kappa)} = 2\,\mathcal{L}_{11}^{(2\kappa-1)}, \tag{7.69}$$

$$\mathcal{L}_{21}^{(\kappa)} + \mathcal{L}_{22}^{(\kappa)} = 2\,\mathcal{L}_{21}^{(2\kappa)}, \tag{7.70}$$

$$\mathcal{L}_{21}^{(\kappa)} - \mathcal{L}_{22}^{(\kappa)} = 2\,\mathcal{L}_{21}^{(2\kappa-1)}. \tag{7.71}$$

Substituting the translations U, V and rotations φ of the mappings (7.60), (7.61), (7.62) and (7.63) into the first set of governing equations (7.53), (7.54) and (7.55). Taking into account eqns. (7.64)–(7.71) we come to the operator representation of the governing equations in terms of the continuous functions $U \equiv U(x_1, x_2, t)$, $V \equiv V(x_1, x_2, t)$ and $\bar{\varphi} \equiv \bar{\varphi}(x_1, x_2, t)$:

$$2\left(\mathcal{L}_{11}^{(2\kappa)} + \vartheta\,\mathcal{L}_{21}^{(2\kappa)}\right) U + 2\,\vartheta\,\mathcal{L}_{21}^{(2\kappa-1)}\,\bar{\varphi} = \left(\frac{\eta}{C}\right)^2 \frac{\partial^2 U}{\partial t^2}, \tag{7.72}$$

$$2\left(\mathcal{L}_{21}^{(2\kappa)} + \vartheta\,\mathcal{L}_{11}^{(2\kappa)}\right) V - 2\,\vartheta\,\mathcal{L}_{11}^{(2\kappa-1)}\,\bar{\varphi} = \left(\frac{\eta}{C}\right)^2 \frac{\partial^2 V}{\partial t^2}, \tag{7.73}$$

$$\mathcal{L}_{21}^{(2\kappa-1)}\,U - \mathcal{L}_{11}^{(2\kappa-1)}\,V + \left(4 + \mathcal{L}_{11}^{(2\kappa)} + \mathcal{L}_{21}^{(2\kappa)}\right)\bar{\varphi}$$

$$= -0.2\,\vartheta^{-1}\left(\frac{\eta}{C}\right)^2 \frac{\partial^2 \bar{\varphi}}{\partial t^2}. \tag{7.74}$$

In view of the relation $\bar{\varphi} = \eta\varphi/2$, these equations become

$$2\left(\mathcal{L}_{11}^{(2\kappa)} + \vartheta\,\mathcal{L}_{21}^{(2\kappa)}\right) U + \vartheta\,\eta\,\mathcal{L}_{21}^{(2\kappa-1)}\,\varphi = \left(\frac{\eta}{C}\right)^2 \frac{\partial^2 U}{\partial t^2}, \tag{7.75}$$

$$2\left(\mathcal{L}_{21}^{(2\kappa)} + \vartheta\,\mathcal{L}_{11}^{(2\kappa)}\right) V - \vartheta\,\eta\,\mathcal{L}_{11}^{(2\kappa-1)}\,\varphi = \left(\frac{\eta}{C}\right)^2 \frac{\partial^2 V}{\partial t^2}, \tag{7.76}$$

$$\mathcal{L}_{21}^{(2\kappa-1)}\,U - \mathcal{L}_{11}^{(2\kappa-1)}\,V + \frac{\eta}{2}\left(4 + \mathcal{L}_{11}^{(2\kappa)} + \mathcal{L}_{21}^{(2\kappa)}\right)\varphi$$

$$= -0.1\left(\frac{\eta}{\vartheta}\right)\left(\frac{\eta}{C}\right)^2 \frac{\partial^2 \bar{\varphi}}{\partial t^2}. \tag{7.77}$$

Substituting the differential operators $\mathcal{L}_{11}^{(2\kappa)}$, $\mathcal{L}_{21}^{(2\kappa)}$, $\mathcal{L}_{11}^{(2\kappa-1)}$, $\mathcal{L}_{21}^{(2\kappa-1)}$ from eqns. (7.68)–(7.71) into eqns. (7.75), (7.76) and (7.77), and allowing for eqns. (7.64)–(7.67), we obtain

$$\frac{\partial^2 U}{\partial x_1^2} + \vartheta\left(\frac{\partial^2 U}{\partial x_2^2} + \frac{\partial \varphi}{\partial x_2}\right) + \mathcal{O}(\eta^2) = C^{-2}\,\frac{\partial^2 U}{\partial t^2}, \tag{7.78}$$

$$\frac{\partial^2 V}{\partial x_2^2} + \vartheta\left(\frac{\partial^2 V}{\partial x_1^2} - \frac{\partial \varphi}{\partial x_1}\right) + \mathcal{O}(\eta^2) = C^{-2}\,\frac{\partial^2 V}{\partial t^2}, \tag{7.79}$$

$$\frac{\partial U}{\partial x_2} - \frac{\partial V}{\partial x_1} + 2\,\varphi + \mathcal{O}(\eta^2) = -\frac{\eta^2}{10}\,\vartheta^{-1}\,C^{-2}\,\frac{\partial^2 U}{\partial t^2} \tag{7.80}$$

where $\mathcal{O}(\eta^2)$ denotes the terms of order η^2 and higher. In the continuum approximation $\eta \to 0$ and, hence, all the components $\mathcal{O}(\eta^2)$ disappear. In addition, the right-hand side of eqn. (7.80) also goes to zero. Dropping these terms and eliminating φ we reduce eqns. (7.78)–(7.80) to

$$\frac{\partial^2 U}{\partial x_1^2} + \frac{\vartheta}{2}\frac{\partial^2 U}{\partial x_2^2} + \frac{\vartheta}{2}\frac{\partial^2 V}{\partial x_1 \, \partial x_2} = C^{-2}\frac{\partial^2 U}{\partial t^2}, \tag{7.81}$$

$$\frac{\partial^2 V}{\partial x_2^2} + \frac{\vartheta}{2}\frac{\partial^2 V}{\partial x_1^2} + \frac{\vartheta}{2}\frac{\partial^2 U}{\partial x_1 \, \partial x_2} = C^{-2}\frac{\partial^2 V}{\partial t^2}, \tag{7.82}$$

We can now compare these differential equations with the similar Navier equations of an elastic, homogeneous, isotropic continuum (Mal and Singh 1991). In our case of the plain problem (plain strain) Navier's equations assume the form

$$\frac{\partial^2 U}{\partial x_1^2} + \frac{\mu}{\lambda+2\mu}\frac{\partial^2 U}{\partial x_2^2} + \frac{\lambda+\mu}{\lambda+2\mu}\frac{\partial^2 V}{\partial x_1 \, \partial x_2} = C_\circ^{-2}\frac{\partial^2 U}{\partial t^2}, \tag{7.83}$$

$$\frac{\partial^2 V}{\partial x_2^2} + \frac{\mu}{\lambda+2\mu}\frac{\partial^2 V}{\partial x_1^2} + \frac{\lambda+\mu}{\lambda+2\mu}\frac{\partial^2 U}{\partial x_1 \, \partial x_2} = C_\circ^{-2}\frac{\partial^2 V}{\partial t^2}, \tag{7.84}$$

where λ and μ are the Lamé constants, $C_\circ$ is the velocity of compressional (or P) wave defined by

$$C_\circ = \sqrt{\frac{(\lambda+2\mu)}{\rho}}. \tag{7.85}$$

It is clear that the two sets of equations—both (7.81), (7.82) and (7.83), (7.84)—have identical structure. They coincide if their coefficients are related by

$$\frac{\vartheta}{2} = \frac{\mu}{\lambda+2\mu}, \tag{7.86}$$

$$\frac{\vartheta}{2} = \frac{\lambda+\mu}{\lambda+2\mu}, \tag{7.87}$$

$$C = C_\circ. \tag{7.88}$$

The equality (7.88) attaches a definite physical meaning to the parameter C introduced earlier by the micromechanical relation (7.59): C is the velocity of the P-wave in an elastic, homogeneous, isotropic continuum. Equations (7.86) and (7.87) yield

$$\lambda = 0, \tag{7.89}$$

$$\vartheta = 1. \tag{7.90}$$

In view of eqns. (7.59), (7.85) and (7.89), eqn. (7.88) yields

$$A_\circ = 2\mu. \tag{7.91}$$

This relation, along with eqns. (7.59) and (7.90), means that

$$I_\circ = A_\circ = 2\mu. \tag{7.92}$$

Recalling that Young's modulus E and Poisson's ratio ν are determined by (Mal and Singh 1991)

$$E = \frac{\mu\,(3\lambda + 2\mu)}{\lambda + \mu}, \tag{7.93}$$

$$\nu = \frac{\lambda}{2\,(\lambda + \mu)} \tag{7.94}$$

and comparing eqn. (7.93) with eqns. (7.89) and (7.92), on the one hand, and eqn. (7.94) with eqn. (7.89), on the other hand, we find the elastic constants to be

$$A_\circ = I_\circ = E, \tag{7.95}$$

$$\nu = 0. \tag{7.96}$$

In view of eqns. (7.59) and (7.95), the velocity C is now expressed as

$$C = \sqrt{\frac{E}{\rho}}. \tag{7.97}$$

We thus see that in a continuum approximation of the theory, the present solid of a simple cubic microstructure is characterized by two different elastic micromoduli $A_\circ$ and $I_\circ$. It becomes an isotropic medium as soon as these microconstants assume equal values, i.e., $A_\circ = I_\circ = E$. In such an isotropic continuum, Poisson's ratio ν is equal to zero.

7.8 Steady-State Waves – General Solution

For studying wave propagation in a solid we will apply the difference governing eqns. (7.53)–(7.59). Some relations of the previous chapter will also be referred to. With regard to eqn. (7.90), the governing equations take the form:

1) For the particles inside the solid ($q = 1, 2, \ldots$)

$$\begin{aligned} &U(p+1,q) - 4U(p,q) + U(p-1,q) + U(p,q+1) + U(p,q-1) \\ &\quad + \bar{\varphi}\,(p,q+1) - \bar{\varphi}(p,q-1) = \left(\frac{\eta}{C}\right)^2 \frac{d^2U(p,q)}{dt^2}, \end{aligned} \tag{7.98}$$

$$\begin{aligned} &V(p,q+1) - 4V(p,q) + V(p,q-1) + V(p+1,q) + V(p-1,q) \\ &\quad - \bar{\varphi}(p+1,q) + \bar{\varphi}(p-1,q) = \left(\frac{\eta}{C}\right)^2 \frac{d^2V(p,q)}{dt^2}, \end{aligned} \tag{7.99}$$

$$U(p,q+1)-U(p,q-1)-V(p+1,q)+V(p-1,q)+4\bar{\varphi}(p,q)$$
$$+\bar{\varphi}(p+1,q)+\bar{\varphi}(p-1,q)+\bar{\varphi}(p,q+1)+\bar{\varphi}(p,q-1)$$
$$=-0.4\left(\frac{\eta}{C}\right)\frac{d^2\bar{\varphi}(p,q)}{dt^2}. \quad (7.100)$$

2) For the particles on the plane boundary of the solid ($q=0$)

$$U(p+1,0)-3U(p,0)+U(p-1,0)+U(p,1)+\bar{\varphi}(p,0)+\bar{\varphi}(p,1)$$
$$=\left(\frac{\eta}{C}\right)^2\frac{d^2U(p,0)}{dt^2}, \quad (7.101)$$

$$V(p,1)-3V(p,0)+V(p+1,0)+V(p-1,0)$$
$$-\bar{\varphi}(p+1,0)+\bar{\varphi}(p-1,0)=\left(\frac{\eta}{C}\right)^2\frac{d^2V(p,0)}{dt^2}, \quad (7.102)$$

$$U(p,1)-U(p,0)-V(p+1,0)+V(p-1,0)+3\bar{\varphi}(p,0)+\bar{\varphi}(p+1,0)$$
$$+\bar{\varphi}(p-1,0)+\bar{\varphi}(p,1)=-0.4\left(\frac{\eta}{C}\right)^2\frac{d^2\bar{\varphi}(p,0)}{dt^2} \quad (7.103)$$

We seek a solution of eqns. (7.98), (7.99), and (7.100) in the form of a harmonic, or *steady-state*, wave propagating in the x_1- or p-direction along the free surface of the half-infinite solid (Fig. 7.1):

$$\left\{\begin{array}{l} U(p,q;t) \\ V(p,q;t) \\ \bar{\varphi}(p,q;t) \end{array}\right\}=\left\{\begin{array}{l} U(q) \\ V(q) \\ \bar{\varphi}(q) \end{array}\right\}\exp[i(k\eta p-\omega t)]. \quad (7.104)$$

Here, and hereafter, $i=\sqrt{-1}$, $k=2\pi/l$; k, l and ω are the wave number, wavelength and circular frequency, respectively. Equation (7.104) results in

$$\left\{\begin{array}{l} U(p+1,q;t) \\ V(p+1,q;t) \\ \bar{\varphi}(p+1,q;t) \end{array}\right\}=\left\{\begin{array}{l} U(q) \\ V(q) \\ \bar{\varphi}(q) \end{array}\right\}\exp\{i[k\eta\,(p+1)-\omega t]\}. \quad (7.105)$$

$$\left\{\begin{array}{l} U(p-1,q;t) \\ V(p-1,q;t) \\ \bar{\varphi}(p-1,q;t) \end{array}\right\}=\left\{\begin{array}{l} U(q) \\ V(q) \\ \bar{\varphi}(q) \end{array}\right\}\exp\{i[k\eta\,(p-1)-\omega t]\}. \quad (7.106)$$

$$\left\{\begin{array}{l} U(p,q+1;t) \\ V(p,q+1;t) \\ \bar{\varphi}(p,q+1;t) \end{array}\right\}=\left\{\begin{array}{l} U(q+1) \\ V(q+1) \\ \bar{\varphi}(q+1) \end{array}\right\}\exp[i(k\eta p-\omega t)]. \quad (7.107)$$

$$\left\{\begin{array}{l} U(p,q-1;t) \\ V(p,q-1;t) \\ \bar{\varphi}(p,q-1;t) \end{array}\right\}=\left\{\begin{array}{l} U(q-1) \\ V(q-1) \\ \bar{\varphi}(q-1) \end{array}\right\}\exp[i(k\eta p-\omega t)]. \quad (7.108)$$

Substitution of eqns. (7.104)–(7.108) into eqns. (7.98), (7.99) and (7.100) gives the following equations in terms of $U(q)$, $U(q+1)$, $U(q-1)$, and so forth:

$$U(q+1)+a_1\,U(q)+U(q-1)+\bar{\varphi}(q+1)-\bar{\varphi}(q-1)=0, \tag{7.109}$$

$$V(q+1)+a_1\,V(q)+V(q-1)-i\,a_2\,\bar{\varphi}(q)=0, \tag{7.110}$$

$$U(q+1)-U(q-1)-i\,a_2\,V(q)+\bar{\varphi}(q+1)+a_3\,\bar{\varphi}(q)+\bar{\varphi}(q-1)=0. \tag{7.111}$$

Similarly, eqns. (7.101), (7.102), (7.103) become

$$U(1)+(1+a_1)\,U(0)+\bar{\varphi}(0)+\bar{\varphi}(1)=0, \tag{7.112}$$

$$V(1)+(1+a_1)V(0)-i\,a_2\,\bar{\varphi}(0)=0, \tag{7.113}$$

$$U(1)-U(0)-i\,a_2\,V(0)+(a_3-1)\bar{\varphi}(0)+\bar{\varphi}(1)=0 \tag{7.114}$$

where

$$a_1=2\,\cos\psi-4+\xi^2\,\psi^2,\ a_2=2\,\sin\psi,\ a_3=2\,\cos\psi+4+0.4\,\xi^2\psi^2 \tag{7.115}$$

$$\psi=k\eta=\frac{2\pi}{\beta},\quad \beta=\frac{l}{\eta},\quad \xi=\frac{c}{C},\quad c=\frac{\omega}{k}, \tag{7.116}$$

c being the wave velocity. The relations (7.109)–(7.114) represent a set of homogeneous linear difference equations in q with constant coefficients. To elucidate the technique of solving such equations we proceed according to the theory of difference equations (Brand 1966).

Let us introduce the difference operator $\mathcal{E}$ by

$$\mathcal{E}\,f(q)=f(q+1). \tag{7.117}$$

Repetitions of $\mathcal{E}$ are written as powers; thus,

$$\mathcal{E}^n\,f(q)=f(q+n). \tag{7.118}$$

We also have

$$\mathcal{E}^0\,f(q)=f(q),\qquad \mathcal{E}^{-1}\,f(q)=f(q-1). \tag{7.119}$$

Given these definitions, we can write the difference equations (7.109), (7.110), (7.111) as

$$F[\mathcal{E}]\left\{\begin{array}{c}U(q)\\V(q)\\\bar{\varphi}(q)\end{array}\right\}=\left\{\begin{array}{c}0\\0\\0\end{array}\right\} \tag{7.120}$$

where $F[\mathcal{E}]$ is the operator matrix

$$F[\mathcal{E}]=\left[\begin{array}{ccc}\mathcal{E}^2+a_1\,\mathcal{E}+1 & 0 & \mathcal{E}^2-1\\ 0 & \mathcal{E}^2+a_1\,\mathcal{E}+1 & -i\,a_2\,\mathcal{E}\\ \mathcal{E}^2-1 & -i\,a_2\,\mathcal{E} & \mathcal{E}^2+a_3\,\mathcal{E}+1\end{array}\right] \tag{7.121}$$

The general solution of the homogeneous equations (7.120) may be written as soon as the roots of the characteristic equation $F(\zeta)=0$ are known. Allowing for eqn. (7.121), we must solve the characteristic equation

$$F(\zeta) \equiv \det \begin{bmatrix} \zeta^2 + a_1\,\zeta + 1 & 0 & \zeta^2 - 1 \\ 0 & \zeta^2 + a_1\,\zeta + 1 & -i\,a_2\,\zeta \\ \zeta^2 - 1 & -i\,a_2\,\zeta & \zeta^2 + a_3\,\zeta + 1 \end{bmatrix} = 0. \quad (7.122)$$

Equation (7.122) yields the two algebraic equations

$$\zeta^2 + a_1\,\zeta + 1 = 0, \quad (7.123)$$

$$\zeta^2 + b_1\,\zeta + 1 = 0 \quad (7.124)$$

in which

$$b_1 = \frac{4 + a_1\,a_3 + a_2^2}{a_1 + a_3}. \quad (7.125)$$

Equations (7.123) and (7.124) have the roots

$$\left\{ \begin{array}{c} \zeta_1 \\ \zeta_3 \end{array} \right\} = \left\{ \begin{array}{c} -a_1/2 - \sqrt{D_1} \\ -a_1/2 + \sqrt{D_1} \end{array} \right\}, \quad (7.126)$$

$$\left\{ \begin{array}{c} \zeta_2 \\ \zeta_4 \end{array} \right\} = \left\{ \begin{array}{c} -b_1/2 - \sqrt{D_2} \\ -b_1/2 + \sqrt{D_2} \end{array} \right\}, \quad (7.127)$$

where

$$D_1 = \frac{a_1^2}{4} - 1, \quad D_2 = \frac{b_1^2}{4} - 1. \quad (7.128)$$

As seen from eqns. (7.126) and (7.128), the roots ζ_1, ζ_3 will be real if $D \geq 0$ or $|a_1| \geq 2$. In view of eqn. (7.115) the latter condition brings about the following restriction imposed on the dimensionless (relative) wave velocity ξ:

$$\xi \leq \xi_P \equiv \frac{\sin(\psi/2)}{\psi/2}. \quad (7.129)$$

The same reasoning for the roots ζ_2, ζ_4 on the basis of eqns. (7.127) and (7.128) results in another limitation on the velocity ξ:

$$\xi \leq \xi_S \equiv \psi^{-1} \sqrt{b_2 - \sqrt{b_2^2 - b_3}} \quad (7.130)$$

in which the parameters b_2 and b_3 are

$$b_2 = 7.5 + 1.5\,\cos\psi + 1, \qquad b_3 = 20\,(1 - \cos\psi). \quad (7.131)$$

The quantities $\xi_{\rm P}$, $\xi_{\rm S}$ may be defined as $\xi_{\rm P} \equiv C_{\rm P}/C$, $\xi_{\rm S} \equiv C_{\rm S}/C$, where $C_{\rm P}$ and $C_{\rm S}$ are the velocities of respectively P and S (shear) plane waves in the solid with microstructure. These statements will be proven below. According to eqns. (7.129), (7.130) and (7.131) both $\xi_{\rm P}$ and $\xi_{\rm S}$ depend on the parameter $\psi = 2\pi/\eta$ and, thus, are functions of the relative wavelength $\beta \equiv l/\eta$: $\xi_{\rm P} = \xi_{\rm P}(\beta)$, $\xi_{\rm S} = \xi_{\rm S}(\beta)$. This means that P- and S-waves are *microscopically* dispersive.

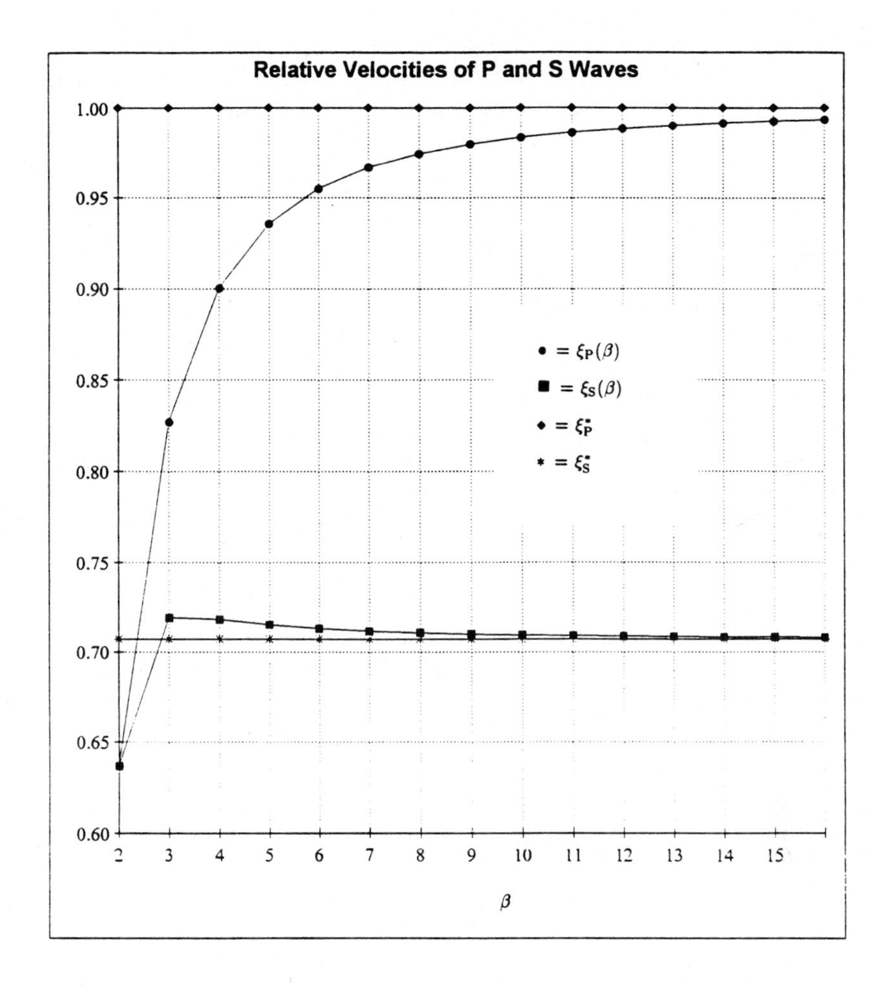

Fig. 7.6. Relative velocities of P and S waves

The functions $\xi_{\mathrm{P}}(\beta)$ and $\xi_{\mathrm{S}}(\beta)$ are shown in Fig. 7.6. As seen from the figure, at $\beta \equiv l/\eta \to \infty$ the velocities $\xi_{\mathrm{P}}(\beta)$ and $\xi_{\mathrm{S}}(\beta)$ tend to certain limits: $\xi_{\mathrm{P}}(\beta) \to \xi_{\mathrm{P}}^* = 1$ and $\xi_{\mathrm{S}}(\beta) \to \xi_{\mathrm{S}}^* = \sqrt{1/2} \doteq 0.707107$ (the exact values are obtained in terms of eqns. (7.129), (7.130) and (7.131)). For any wavelength $l > 0$, the limit $l/\eta \to \infty$ is reached at $\eta \to 0$, i.e., in a continuum approximation. It follows that ξ_{P}^* and ξ_{S}^* are ultimate dimensionless velocities of P- and S-waves in an unbounded linear-elastic, homogeneous, isotropic continuum (ξ_{S}^* holds only with Poisson's ratio $\nu = 0$).

The above microscopic dispersion is mainly pronounced over the bands of relatively short wavelengths (Fig. 7.6):

- the velocities $\xi_{\mathrm{S}}(\beta)$ depart from their continuum counterpart $\xi_{\mathrm{S}}^* = 0.707107$ by 1% and more at the relative wavelengths $2 < \beta \equiv l/\eta < 5$,
- the velocities $\xi_{\mathrm{P}}(\beta)$ depart from their continuum limit $\xi_{\mathrm{P}}^* = 1$ by 1% and more over the wave band $2 < \beta < 12$ which is wider than that for S waves.

We now return to the relations (7.126) and (7.127). Assuming that the restrictions (7.129) and (7.130) are valid, we get four distinct real roots η_r, $r = 1, 2, 3, 4$, that generate four independent solutions of the difference equations (7.120)

$$\left\{ \begin{array}{c} U_r(q) \\ V_r(q) \\ \bar{\varphi}_r(q) \end{array} \right\} = \left[\begin{array}{ccc} A_r & 0 & 0 \\ 0 & B_r & 0 \\ 0 & 0 & C_r \end{array} \right] \left\{ \begin{array}{c} \zeta_r^q \\ \zeta_r^q \\ \zeta_r^q \end{array} \right\} \tag{7.132}$$

Here $U_r(q)$, $V_r(q)$, $\bar{\varphi}(q)$ are the amplitudes of an r-mode of the wave at an arbitrary depth of the solid $x_2 > 0$ or $q \geq 1$; A_r, B_r, C_r being the same amplitudes on the free surface of the solid $x_2 = 0$ or $q = 0$ (Fig. 7.1). Thus, any solution (7.132) describes an r-mode of the steady-state wave (7.104) whose amplitudes either attenuate (if $\zeta_r < 1$) or grow (if $\zeta > 1$) with the depth q without bound.

Analysis of the real roots ζ_r in terms of eqns. (7.126) and (7.127) and the restrictions (7.129) and (7.130) show that $\zeta_1 < 1$ and $\zeta_2 < 1$, whereas $\zeta_3 > 1$ and $\zeta_4 > 1$. Since there are no energy sources inside the solid the growing modes 3 and 4 are physical impossibilities and have therefore to be neglected. As a result, eqns. (7.132) reduce to the two attenuating modes 1 and 2:

$$\left\{ \begin{array}{c} U_1(q) \\ V_1(q) \\ \bar{\varphi}_1(q) \end{array} \right\} = \left[\begin{array}{ccc} A_1 & 0 & 0 \\ 0 & B_1 & 0 \\ 0 & 0 & C_1 \end{array} \right] \left\{ \begin{array}{c} \zeta_1^q \\ \zeta_1^q \\ \zeta_1^r \end{array} \right\} \tag{7.133}$$

$$\left\{ \begin{array}{c} U_2(q) \\ V_2(q) \\ \bar{\varphi}_2(q) \end{array} \right\} = \left[\begin{array}{ccc} A_2 & 0 & 0 \\ 0 & B_2 & 0 \\ 0 & 0 & C_2 \end{array} \right] \left\{ \begin{array}{c} \zeta_2^q \\ \zeta_2^q \\ \zeta_2^q \end{array} \right\} \tag{7.134}$$

where the parameters of attenuation ζ_1 and ζ_2 are defined by eqns. (7.126) and (7.127) respectively. Allowance must also be made for eqns. (7.115) and (7.116). The relationships (7.133) and (7.134) represent a general solution for the steady-state wave *inside* the solid where the discrete vertical coordinate $q \geq 1$.

It should be noted that eqns. (7.133) and (7.134) contain two unknown discrete functions ζ_1 and ζ_2. In view of eqns. (7.126) and (7.127) and eqns. (7.115) and (7.116), ζ_1 and ζ_2 depend not only on the relative wavelength $\beta \equiv l/\eta$ but also on the third unknown function, the relative velocity of the wave $\xi \equiv c/C$. This velocity, too, depends on β. Thus, we have three unknown functions $\zeta_1(\beta)$, $\zeta_2(\beta)$, $\xi(\beta)$ of the independent variable β that cannot be unambiguously calculated in terms of only two algebraic equations (7.126) and (7.127). There must be a third equation. Such an additional equation can be derived by applying the boundary conditions (7.112)–(7.114). Before solving the problem, we will briefly touch upon P- and S-waves.

7.9 Steady-State P- and S-Waves

The above consideration has concerned waves with attenuation whose parameters $\zeta_1 < 1$ and $\zeta_2 < 1$ are supposed to be unknown beforehand and must therefore be calculated by using eqns. (7.123) and (7.124). There is, however, another approach. We may consider simpler steady-state waves, those without attenuation. For such waves, the parameters ζ_1 and ζ_2 cease to be unknown because they are taken beforehand as unity: $\zeta_1 = \zeta_2 = 1$. Then eqns. (7.123) and (7.124) become

$$a_1 + 2 = 0, \tag{7.135}$$

$$b_1 + 2 = 0, \tag{7.136}$$

and eqns. (7.133) and (7.134) reduce to

$$\left\{ \begin{array}{c} U(q) \\ V(q) \\ \bar{\varphi}(q) \end{array} \right\} = \left\{ \begin{array}{c} \tilde{A} \\ \tilde{B} \\ \tilde{C} \end{array} \right\} \tag{7.137}$$

where $\tilde{A}$, $\tilde{B}$, $\tilde{C}$ are constants. According to eqn. (7.137), the difference operator $\mathcal{E}$ by eqn. (7.117) becomes unity ($\mathcal{E} = 1$). Hence the operator matrix (7.121) is transformed into

$$F[1] = \left[\begin{array}{ccc} a_1 + 2 & 0 & 0 \\ 0 & a_1 + 2 & -i\,a_2 \\ 0 & -i\,a_2 & a_3 + 2 \end{array} \right] \tag{7.138}$$

In view of eqns. (7.137) and (7.138) the governing equations (7.120) take the form of homogeneous equations for the constants $\tilde{A}$, $\tilde{B}$, $\tilde{C}$

$$\left[\begin{array}{ccc} a_1 + 2 & 0 & 0 \\ 0 & a_1 + 2 & -i\,a_2 \\ 0 & -i\,a_2 & a_3 + 2 \end{array} \right] \left\{ \begin{array}{c} \tilde{A} \\ \tilde{B} \\ \tilde{C} \end{array} \right\} = \left\{ \begin{array}{c} 0 \\ 0 \\ 0 \end{array} \right\} \tag{7.139}$$

or, equivalently,

$$(a_1 + 2)\,\tilde{A} = 0, \tag{7.140}$$

$$\begin{bmatrix} a_1 + 2 & -i\,a_2 \\ -i\,a_2 & a_3 + 2 \end{bmatrix} \begin{Bmatrix} \tilde{B} \\ \tilde{C} \end{Bmatrix} = \begin{Bmatrix} 0 \\ 0 \end{Bmatrix}. \tag{7.141}$$

Equations (7.139)–(7.141) can also be obtained without referring to the operator equations (7.120) by substituting the assumed solution (7.137) into the governing equations (7.109), (7.110) and (7.111) written in a conventional, non-operator form. Due to eqn. (7.137), the sets of equations (7.140) and (7.141) determine two independent waves: the first wave with $U(q) = \tilde{A}$, $V(q) = 0$, $\bar{\varphi}(q) = 0$, and the second wave with $U(q) = 0$, $V(q) = \tilde{B}$, $\bar{\varphi}(q) = \tilde{C}$. Let us examine these cases.

1. Equation (7.140) has a nontrivial solution for $\tilde{A}$ if $a_1 + 2 = 0$, i.e., when eqn. (7.135) holds. Solving eqn. (7.135) we come to the wave velocity ξ_P by eqn. (7.129). We have already mentioned that ξ_P is the velocity of P wave. Now the statement can easily be proven. Indeed, the only displacements $U(q) = \tilde{A}$ do not attenuate with the depth q and are parallel to the p-direction along which the plane wave propagates (see eqn. (7.104)). These properties determine a longitudinal, or compressional, or P-wave.
2. Equations (7.141) have nontrivial solutions for $\tilde{B}$ and $\tilde{C}$ if

$$\det \begin{bmatrix} a_1 + 2 & -i\,a_2 \\ -i\,a_2 & a_3 + 2 \end{bmatrix} = 0. \tag{7.142}$$

Given the relation (7.125), eqn. (7.142) is identical to eqn. (7.136) whose solution brings about the velocity ξ_S according to eqn. (7.130). This velocity has been attributed to S-waves in advance. To prove this we take into account that the remained translations $V(q) = \tilde{B}$ and rotations $\bar{\varphi}(q) = \tilde{C}$ do not attenuate with the depth q, and that $V(q)$ is perpendicular to the p-direction of wave propagation. These properties are the salient features of S-waves. It should be added that unlike P-waves, S-waves cannot travel along the free surface of a half-infinite solid; they do not satisfy the boundary conditions (7.112), (7.113), (7.114) in view of eqn. (7.141). S-waves can propagate in any direction only in unbounded media.

7.10 Steady-State Surface Wave

We now seek a steady-state wave different from a P-wave but which might, nonetheless, be able to travel along the free boundary of the present half-infinite solid. The features of P- and S-waves suggest that a presumable solution may be found only in terms of waves with attenuation. Thus, we return to the attenuating modes 1 and 2 defined by eqns. (7.133) and (7.134) and assume the solution to be their linear combination:

$$\left\{\begin{array}{c} U(q) \\ V(q) \\ \bar{\varphi}(q) \end{array}\right\} = \left\{\begin{array}{c} U_1(q) \\ V_1(q) \\ \bar{\varphi}_1(q) \end{array}\right\} + \left\{\begin{array}{c} U_2(q) \\ V_2(q) \\ \bar{\varphi}_2(q) \end{array}\right\} = \left[\begin{array}{cc} \tilde{A}_1 & \tilde{A}_2 \\ \tilde{B}_1 & \tilde{B}_2 \\ \tilde{C}_1 & \tilde{C}_2 \end{array}\right] \left\{\begin{array}{c} \zeta_1^q \\ \zeta_2^q \end{array}\right\} \tag{7.143}$$

The parameters of attenuation ζ_1 and ζ_2 in eqn. (7.143) are now determined by eqns. (7.123) and (7.124), respectively, or

$$\zeta_1^2 + a_1\,\zeta_1 + 1 = 0, \tag{7.144}$$

$$\zeta_2^2 + b_1\,\zeta_2 + 1 = 0. \tag{7.145}$$

Upon substituting $U(q)$, $V(q)$, $\bar{\varphi}(q)$ from eqn. (7.143) into the governing equations (7.109), (7.110) and (7.111) for the particles *inside* the solid, we obtain $(q = 1, 2, \ldots)$

$$\begin{aligned} &\zeta_1^{q-1}\left(\tilde{A}_1\,(\zeta_1^2 + a_1\,\zeta_1 + 1) + \tilde{C}_1\,(\zeta_1^2 - 1)\right) \\ &\quad + \zeta_1^{q-1}\left(\tilde{A}_2\,(\zeta_2^2 + a_1\,\zeta_2 + 1) + \tilde{C}_2\,(\zeta_2^2 - 1)\right) = 0, \end{aligned} \tag{7.146}$$

$$\begin{aligned} &\zeta_1^{q-1}\left(\tilde{B}_1\,(\zeta_1^2 + a_1\,\zeta_1 + 1) - i\,\tilde{C}_1\,a_2\,\zeta_1\right) \\ &\quad + \zeta_1^{q-1}\left(\tilde{B}_2\,(\zeta_2^2 + a_1\,\zeta_2 + 1) - i\,\tilde{C}_2\,a_2\,\zeta_2\right) = 0, \end{aligned} \tag{7.147}$$

$$\begin{aligned} &\zeta_1^{q-1}\left(\tilde{A}_1\,(\zeta_1^2 - 1) - i\,\tilde{B}_1\,a_2\,\zeta_1 + \tilde{C}_1\,(\zeta_1^2 + a_3\,\zeta_1 + 1)\right) \\ &\quad + \zeta_2^{q-1}\left(\tilde{A}_1\,(\zeta_2^2 - 1) - i\,\tilde{B}_2\,a_2\,\zeta_2 + \tilde{C}_2\,(\zeta_2^2 + a_3\,\zeta_2 + 1)\right) = 0. \end{aligned} \tag{7.148}$$

Since $\zeta_1 \neq \zeta_2$, these equations are satisfied at any $q = 1, 2, \ldots$ if all the expressions in the large round brackets are equal to zero. Thus, we come to the two new sets of independent homogeneous algebraic equations:

$$\begin{aligned} &\left[\begin{array}{ccc} \zeta_1^2 + a_1\,\zeta_1 + 1 & 0 & \zeta_1^2 - 1 \\ 0 & \zeta_1^2 + a_1\,\zeta_1 + 1 & -i\,a_2\,\zeta_1 \\ \zeta_1^2 - 1 & -i\,a_2\,\zeta_1 & \zeta_1^2 + a_3\,\zeta_1 + 1 \end{array}\right] \left\{\begin{array}{c} \tilde{A}_1 \\ \tilde{B}_1 \\ \tilde{C}_1 \end{array}\right\} \\ &\quad = \left\{\begin{array}{c} 0 \\ 0 \\ 0 \end{array}\right\}, \end{aligned} \tag{7.149}$$

$$\begin{aligned} &\left[\begin{array}{ccc} \zeta_2^2 + a_1\,\zeta_2 + 1 & 0 & \zeta_2^2 - 1 \\ 0 & \zeta_2^2 + a_1\,\zeta_2 + 1 & -i\,a_2\,\zeta_2 \\ \zeta_2^2 - 1 & -i\,a_2\,\zeta_2 & \zeta_2^2 + a_3\,\zeta_2 + 1 \end{array}\right] \left\{\begin{array}{c} \tilde{A}_2 \\ \tilde{B}_2 \\ \tilde{C}_2 \end{array}\right\} \\ &\quad = \left\{\begin{array}{c} 0 \\ 0 \\ 0 \end{array}\right\}, \end{aligned} \tag{7.150}$$

Solving these equations in view of eqns. (7.144) and (7.145), we obtain

$$\tilde{B}_1 = -i\,\tilde{A}_1\,\frac{\zeta_1^2 - 1}{a_2\,\zeta_1}, \tag{7.151}$$

$$\tilde{C}_1 = 0, \tag{7.152}$$

$$\tilde{A}_2 = i\,\tilde{B}_2\,\frac{\zeta_2^2 - 1}{a_2\,\zeta_2}, \tag{7.153}$$

$$\tilde{C}_2 = -i\,\tilde{B}_2\,\frac{\zeta_2^2 + a_1\,\zeta_2 + 1}{a_2\,\zeta_2} \tag{7.154}$$

Due to eqns. (7.149) and (7.150), the number of unknown constants is reduced from six ($\tilde{A}_1, \tilde{B}_1, \tilde{C}_1, \tilde{A}_2, \tilde{B}_2, \tilde{C}_2$) to two ($\tilde{A}_1, \tilde{B}_2$). Substituting eqns. (7.151)–(7.154) into eqn. (7.143), we come to the solution in terms of the two constants $\tilde{A}_1$ and $\tilde{B}_2$

$$\left\{ \begin{array}{c} U(q) \\ V(q) \\ \bar{\varphi}(q) \end{array} \right\} = [\mathcal{D}] \left\{ \begin{array}{c} \tilde{A}_1\,\zeta_1^q \\ \tilde{B}_2\,\zeta_2^q \end{array} \right\} \tag{7.155}$$

where

$$[\mathcal{D}] = \left[\begin{array}{cc} 1 & i\,(\zeta_2^2 - 1)\,/\,(a_2\,\zeta_2) \\ -i\,(\zeta_1 - 1)/(a_2\,\zeta_1) & 1 \\ 0 & -i\,(\zeta_2^2 + a_1\,\zeta_2 + 1)/(a_2\,\zeta_2) \end{array} \right] \tag{7.156}$$

It is worth reminding the reader that all the above relations—from eqn. (7.144) to eqn. (7.156)—stem from the governing equations that are valid inside the solid. Now we turn to the governing equations (7.112)–(7.114) for the particles on the boundary of the solid. For convenience, we rewrite the equations as

$$[\mathcal{D}_0] \left[\begin{array}{c} U(0) \\ V(0) \\ \bar{\varphi}(0) \end{array} \right] + [\mathcal{D}_1] \left[\begin{array}{c} U(1) \\ V(1) \\ \bar{\varphi}(1) \end{array} \right] = [\mathcal{Z}] \equiv \left[\begin{array}{c} 0 \\ 0 \\ 0 \end{array} \right]. \tag{7.157}$$

Here

$$[\mathcal{D}_0] = \left[\begin{array}{ccc} a_1 + 1 & 0 & 1 \\ 0 & a_1 + 1 & -i\,a_2 \\ -1 & -i\,a_2 & a_3 - 1 \end{array} \right], \quad [\mathcal{D}_1] = \left[\begin{array}{ccc} 1 & 0 & 1 \\ 0 & 1 & 0 \\ 1 & 0 & 1 \end{array} \right]. \tag{7.158}$$

On the basis of eqn. (7.155), eqns. (7.157) become

$$[\mathcal{D}_0]\,[\mathcal{D}] \left\{ \begin{array}{c} \tilde{A}_1 \\ \tilde{B}_2 \end{array} \right\} + [\mathcal{D}_1]\,[\mathcal{D}] \left\{ \begin{array}{c} \tilde{A}_1\,\zeta_1 \\ \tilde{B}_2\,\zeta_2 \end{array} \right\} = \{\mathcal{Z}\}. \tag{7.159}$$

Multiplying the matrices $[\mathcal{D}_0]\,[\mathcal{D}]$ and $[\mathcal{D}_1]\,[\mathcal{D}]$ according to eqns. (7.156) and (7.158), we bring eqns. (7.159) into the final form

$$g_{11}\,\tilde{A}_1 + g_{12}\,\tilde{B}_2 = 0, \tag{7.160}$$

$$g_{21}\,\tilde{A}_1 + g_{22}\,\tilde{B}_2 = 0, \tag{7.161}$$

$$g_{31}\,\tilde{A}_1 + g_{32}\,\tilde{B}_2 = 0, \tag{7.162}$$

where

$$g_{11} = \zeta_1 + a_1 + 1, \tag{7.163}$$

$$g_{21} = -\frac{i}{a_2\,\zeta_1}(\zeta_1 + a_1 + 1)(\zeta_1^2 - 1), \tag{7.164}$$

$$g_{31} = -\frac{\zeta_1 - 1}{\zeta_1}, \tag{7.165}$$

$$g_{12} = \frac{i}{a_2\,\zeta_2}\left[(\zeta_2 + a_1 + 1)(\zeta_2^2 - 1) - (\zeta_2 + 1)(\zeta_2^2 + a_1\,\zeta_2 + 1)\right], \tag{7.166}$$

$$g_{22} = -\frac{1}{\zeta_2}\left[\zeta_2^2 + a_1\,\zeta_2 + 1 - (\zeta_2 + a_1 + 1)\,\zeta_2\right], \tag{7.167}$$

$$g_{32} = -\frac{i}{a_2\,\zeta_2}\left[(\zeta_2^2 - 1)(\zeta_2 - 1) - a_2^2\,\zeta_2\right. \tag{7.168}$$

$$\left. - (\zeta_2 + a_3 - 1)(\zeta_2^2 + a_1\,\zeta_2 + 1)\right]. \tag{7.169}$$

We have obtained a specific system of three equations (7.160), (7.161) and (7.162) for only two unknown constants $\tilde{A}_1$ and $\tilde{B}_2$. To deal with the system, we must take into account its essential feature. Suppose that we have only two equations (7.160) and (7.161). Such a system has a nontrivial solution for $\tilde{A}_1$ and $\tilde{B}_2$ only if its determinant is equal to zero:

$$\begin{vmatrix} g_{11} & g_{12} \\ g_{21} & g_{22} \end{vmatrix} \equiv g_{11}\,g_{22} - g_{21}\,g_{12} = 0. \tag{7.170}$$

We may also consider the other two equations (7.161) and (7.162) and get, similarly,

$$\begin{vmatrix} g_{21} & g_{22} \\ g_{31} & g_{32} \end{vmatrix} \equiv g_{21}\,g_{32} - g_{31}\,g_{22} = 0. \tag{7.171}$$

At last, we can choose eqns. (7.160) and (7.162) and, thus, come to the third characteristic equation:

$$\begin{vmatrix} g_{11} & g_{12} \\ g_{31} & g_{32} \end{vmatrix} \equiv g_{32}\,g_{11} - g_{12}\,g_{31} = 0. \tag{7.172}$$

In terms of eqns. (7.170) and (7.171) we have

$$g_{11} = \frac{g_{21}\,g_{12}}{g_{22}}, \quad g_{31} = \frac{g_{32}\,g_{21}}{g_{22}}. \tag{7.173}$$

Substituting g_{11} and g_{31} from eqn. (7.173) into eqn. (7.172), we obtain

$$g_{32}\,g_{11} - g_{12}\,g_{31} = g_{32}\left(\frac{g_{21}\,g_{12}}{g_{22}}\right) - g_{12}\left(\frac{g_{32}\,g_{21}}{g_{22}}\right) \equiv 0. \tag{7.174}$$

The identity (7.174) means that there are only two independent characteristic equations, the third one is satisfied identically. Due to this feature, the system of equations (7.160)–(7.162) has a nontrivial solution if any two of the three characteristic equations, for instance eqns. (7.170) and (7.171), are satisfied.

We now return to the main problem of a surface wave that is described by three functions: the two parameters of attenuation $\zeta_1(\beta)$, $\zeta_2(\beta)$ and the relative velocity of the wave $\xi(\beta)$. To compute these functions we have four equations: (7.144), (7.145), on the one hand, and (7.170), (7.171), on the other hand. The system is redundant and must be reduced to three equations. This can be done as follows. Remember that the relations (7.144) and (7.145) stem from the equations of motion *inside* the solid. Therefore they both must be taken into account in *any* wave problem. At the same time, eqns. (7.170) and (7.171) are specific only to the *boundary* problem and are to be added to the main equations (7.144) and (7.145). But because adding them simultaneously is impossible, we are forced to join them one by one. Thus, we come perforce to the following two variants of relations determining a possible surface wave:

- Variant 1 made up of eqns. (7.144), (7.145) and (7.170),
- Variant 2 made up of eqns. (7.144), (7.145) and (7.171).

Denote the sets of solutions $\zeta_1(\beta)$, $\zeta_2(\beta)$, $\xi(\beta)$ stemming from Variants 1 and 2 by $\mathcal{I}_1$ and $\mathcal{I}_2$, respectively. If these sets intersect, i.e., $\mathcal{I} = \mathcal{I}_1 \cap \mathcal{I}_2 \neq \emptyset$, then the intersection $\mathcal{I}$ contains at least one system of the functions $\{\zeta_1(\beta), \zeta_2(\beta), \xi(\beta)\}$ defining a surface wave.

Variants 1 and 2 were solved numerically. Each of them has yielded only one and the same system $\{\zeta_1(\beta), \zeta_2(\beta), \xi(\beta)\} \in \mathcal{I}_1 = \mathcal{I}_2 = \mathcal{I}$. This system characterizes a new surface wave (NSW).

7.11 Discussion

The parameters of attenuation $\zeta_1(\beta)$, $\zeta_2(\beta)$ and the relative velocity $\xi(\beta)$ of the NSW are shown in Fig. 7.7, along with those $\zeta_{1R}(\beta)$, $\zeta_{2R}(\beta)$, ξ_R of the Rayleigh surface wave in the classical elastic continuum. Before analyzing the plots in Fig. 7.7, we will briefly explain the way the RSW parameters $\zeta_{1R}(\beta)$, $\zeta_{2R}(\beta)$, ξ_R have been obtained.

The displacement field of the RSW is expressed by (Bedford and Drumheller 1994)

$$\begin{aligned} U(x_1, x_2, t) &= \left(i\,k\,A\,e^{-hx_2} + h_s\,B\,e^{-h_s x_2}\right) \exp[i(kx_1 - \omega t)], && (7.175) \\ V(x_1, x_2, t) &= \left(-h\,A\,e^{-hx_2} + i\,k\,B\,e^{-h_s x_2}\right) \exp[i(kx_1 - \omega t)] && (7.176) \end{aligned}$$

where the notation is partially changed to be consistent with that in this chapter. In eqns. (7.175) and (7.176), A and B are constants, attenuation being included by the multipliers $\exp(-h\,x_2)$ and $\exp(-h_s\,x_2)$. The parameters h and h_s are taken initially as

$$h = \sqrt{k^2 - \frac{\omega^2}{C}}, \qquad h_s = \sqrt{k^2 - \frac{\omega^2}{C_s^2}}. \qquad (7.177)$$

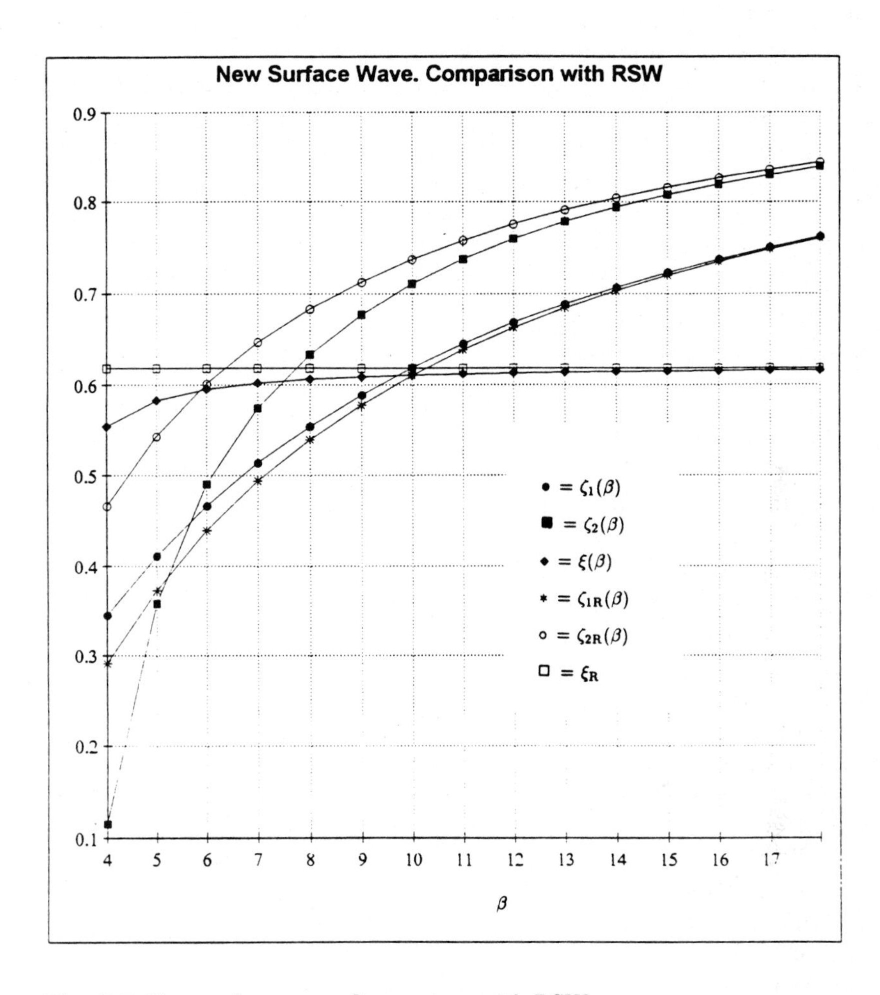

Fig. 7.7. New surface wave. Comparison with RSW

Here C and C_s are the velocities of the P- and S-waves, respectively. If Poisson's ratio $\nu = 0$ (see eqn. (7.96)), then

$$C = \sqrt{\frac{E}{\rho}}, \qquad C_s = \frac{\sqrt{2}}{2} C. \tag{7.178}$$

In view of eqn. (7.178), the expressions (7.177) can be easily transformed into

$$h = k\sqrt{1 - \xi_{\rm R}^2}, \qquad h_s = k\sqrt{1 - 2\,\xi_{\rm R}^2} \tag{7.179}$$

where $\xi_{\rm R} = C_{\rm R}/C = 0.618$ and $C_{\rm R}$ is the velocity of the RSW. Setting $x_2 = q\eta$ (Fig. 7.1) and taking into account eqns. (7.116) and (7.179), we put the above exponents $\exp(-h\,x_2)$ and $\exp(-h_s\,x_2)$ into the form $\exp(-h\,x_2) = \zeta_{1R}^q$, $\exp(-h_s\,x_2) = \zeta_{2\rm R}^q$ which is in agreement with eqn. (7.155), the transformed parameters of attenuation $\zeta_{1\rm R}$, $\zeta_{2\rm R}$ of the RSW now being functions of β:

$$\zeta_{1\rm R}(\beta) = \exp\left(-\frac{2\pi}{\beta}\sqrt{1 - \xi_{\rm R}^2}\right), \tag{7.180}$$

$$\zeta_{2\rm R}(\beta) = \exp\left(-\frac{2\pi}{\beta}\sqrt{1 - 2\xi_{\rm R}^2}\right). \tag{7.181}$$

The variables $\zeta_{1\rm R}(\beta)$, $\zeta_{2\rm R}(\beta)$ of eqns. (7.180) and (7.181) and the constant $\xi_{\rm R} = 0.618$ of the classical RSW are shown in Fig. 7.7 to compare them with the characteristics $\zeta_1(\beta)$, $\zeta_2(\beta)$, $\xi(\beta)$ of the NSW. In addition, Fig. 7.8 plots the ratios $\zeta_1(\beta)/\zeta_{1R}(\beta)$, $\zeta_2(\beta)/\zeta_{2R}(\beta)$ and $\xi(\beta)/\xi_R$ on a percentage basis.

Both Figs. 7.7 and 7.8 display an appreciable departure of the functions $\zeta_1(\beta)$, $\zeta_2(\beta)$, $\xi(\beta)$ from their continuum counterparts $\zeta_{1\rm R}(\beta)$, $\zeta_{2\rm R}(\beta)$, $\xi_{\rm R}$ over the band of relatively short wavelengths $\beta \equiv l/\eta < 18$. The strongest departure is intrinsic in the second parameter of attenuation $\zeta_2(\beta)$, the other two characteristics $\zeta_1(\beta)$ and $\xi(\beta)$ diverge to a lesser extent. Since the dimensionless velocity $\xi(\beta)$ of the NSW has a pronounced dependence on β, the NSW is *microscopically* dispersive. As β increases, the difference between the systems $\{\zeta_1(\beta), \zeta_2(\beta), \xi(\beta)\}$ and $\{\zeta_{1\rm R}(\beta), \zeta_{2\rm R}(\beta), \xi_{\rm R}\}$ gradually diminishes and completely vanishes as $\beta \to \infty$ when the *dispersive* NSW transforms into the *nondispersive* classical RSW.

Thus, whereas the theory of the classical RSW loses its applicability beyond the range of infinitely long waves ($\beta \to \infty$), the theory of the NSW holds for almost the entire Brillouin zone (Brillouin 1963), i.e., from the infinitely long waves ($\beta \to \infty$) down to the waves whose wavelength l is three times the interparticle distance η ($\beta = 3$, not shown in Figs. 7.7 and 7.8). There is only one exception. Unlike P-waves, the NSW cannot propagate at the right edge of the Brillouin zone where $\beta = 2$. It follows that the NSW theory is an extension of the classical theory of the infinitely long RSW traveling in a half-infinite continuum to the surface waves of almost any wavelength that may propagate in a half-infinite discrete microstructure (here, the simple cubic one).

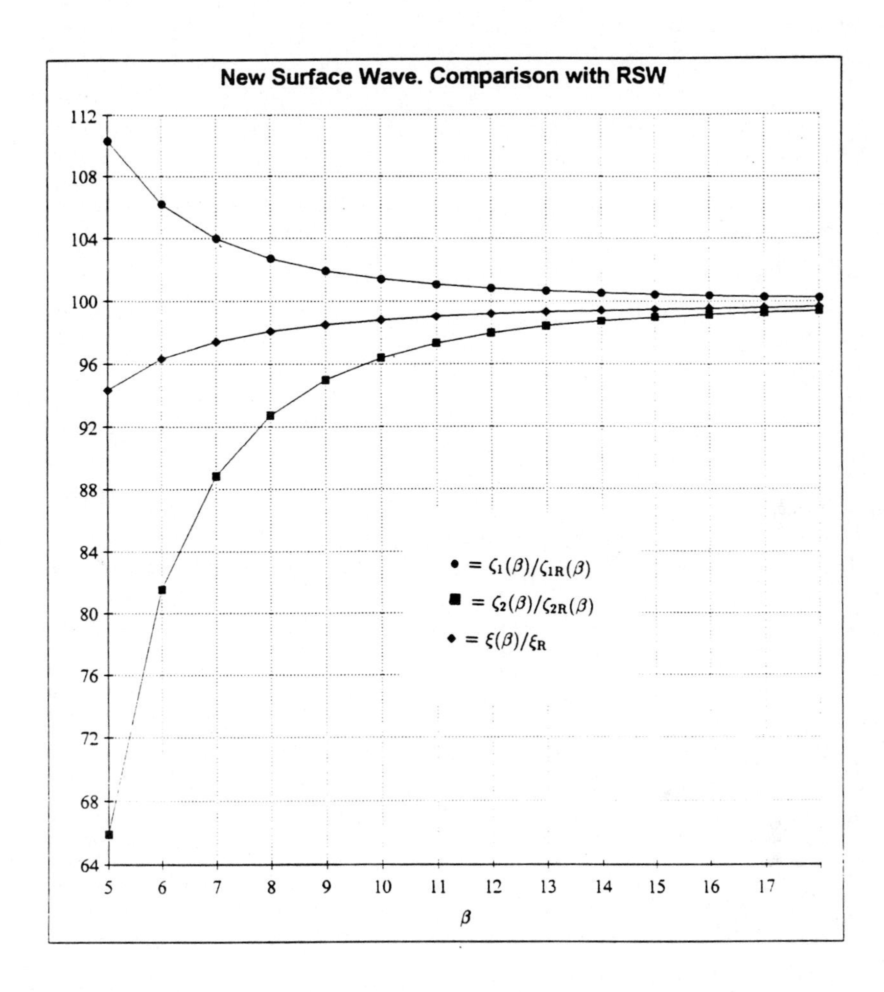

Fig. 7.8. New surface wave. Comparison with RSW on a percentage basis

It is also interesting to compare the microscopic dispersion of the NSW with the macroscopic dispersion of the RSW brought about by the effects of gravity and curvature of a linear-elastic, homogeneous, isotropic solid. The seminal results in the area were established by Bromwich (1898) for incompressible spheres and half-spaces and by Love (1911) for the compressible ones. Later on Sezawa (1927) and Oliver (1955) considered the influence of cylindrical curvature. According to these studies, the phase velocity of the RSW changes as follows:

- due to gravity in the half-space, it increases or decreases (depending on Poisson's ratio) by approximately tenths of a percent,
- owing to curvature, it increases by up to 10%.

Figure 7.8 shows that due to the microscopic dispersion, the phase velocity of the NSW is less than that of the RSW at any wavelength. Over the band of relatively short wavelengths ($18 > \beta > 3$) the decrease ranges from 0.36% to 10.35%. Thus, the effect of microscopic dispersion proves to be comparable with the influence of macroscopic dispersion caused by curvature and is far more perceptible than that induced by gravity.

The problem of the RSW was also touched upon in the monograph by Kunin (1983) in terms of a nonlocal model of a medium with binary (central) interaction characterized by one elastic modulus. In such a medium the Lamé constant $\lambda = \mu$ and consequently Poisson's ratio $\nu = 1/4$. As follows from Sect. 5.4 this particular case corresponds to a solid with a plane hexagonal microstructure. Referring to the so-called "approximation with respect to the first root," the author reports to have obtained "the new surface waves, which are absent in the classical theory" and which "decay in the case of large wavelength, at a distance of the order of the radius of interaction." At the same time the very dispersion law that is to be the main goal of the study and that requires "cumbersome computations" is not obtained.

7.12 Closure

In this chapter we have developed a new theoretical approach to the study of a surface wave propagating in an elastic solid with cubic microstructure. The approach has enabled us to reveal a new surface wave with microscopic dispersion pronounced over the band of relatively short wavelengths. In a continuum approximation the new wave reduces to the conventional nondispersive RSW. Thus, the new surface wave is a generalization of the classical Rayleigh surface wave.

The proposed difference method is mathematically precise. As shown in Sect. 7.7, the classical theory of elasticity is a first (nonscale, nonlocal) approximation of the difference model. The second and further differential approximations include scale parameters to increasing powers and hence are able to approach an exact solution to a sequentially increasing degree of accuracy.

The difference model of DM is evidently simple and can easily be extended to a study of more intricate boundary problems that involve solids with a variety of different microstructures:

1. Hexagonal, b.c.c, f.c.c, etc.
2. Isotropic in a continuum approximation, with Poisson's ratio $\nu \neq 0$.
3. Anisotropic in a continuum approximation.
4. Multilayered.
5. Inhomogeneous geometrically (when parameters of the underlying Bravais lattice are dependent on the lattice coordinates (p, q)).
6. With particles of different types (e.g., "diatomic lattices").
7. With linear but nonlocal and inhomogeneous nearest-neighbor interactions (see Sect. 7.3).
8. With nonlinear nearest-neighbor interactions.
9. With farther than nearest-neighbor interactions, and so on.

Solutions of these new problems are expected to bring about the discovery of new microscopic wave phenomena.

8. Isotropic Plane Elastostatics

J. C. Nadeau, A. H. Nashat and M. Ferrari

8.1 Introduction

We begin with a uniqueness theorem in linear elastic doublet mechanics. With this tool in hand we present two methods of obtaining solutions. The first stems from a correspondence between problems in doublet and continuum mechanics which allows the generation of a solution in one theory given a solution in the other. The second methodology involves combining the micro-stress equilibrium and micro-strain compatibility requirements into a single condition, from which stems a micro-stress function (MSF) analogous to the CM Airy stress function (ASF). While an admissible ASF is any bi-harmonic function, the equation governing the MSF is a more general fourth order differential equation dependent on the lattice geometry. We show that for a specific choice of the DM lattice the MSF a bi-harmonic function.

We illustrate the advantages of the two methodologies by solving several problems in plane elastostatics. We begin by considering homogeneous deformations and then obtain solutions to the classical problems of Flamant and Kelvin. Finally we obtain the stress concentrations due to a circular hole in an infinite plate subjected to bi-axial tension. Our purpose is not to provide a catalog of solutions in DM, but rather to elucidate the techniques.

As an application, Granik and Ferrari (1993) considered the DM equivalent of Flamant's problem: a concentrated force acting normal to the free-surface boundary of a planar elastic half-space. Though the qualitative description of the existing DM solution is accurate, its quantitative inconsistencies are corrected in this chapter.

DM is a scale dependent theory but for the purposes of this chapter we consider only the non-scale subcase.

8.2 Uniqueness Theorem

The objective of this section is to establish a uniqueness theorem in linear elastic DM. Consider a DM body $\mathcal{B}$ with boundary $\partial\mathcal{B}$. The body $\mathcal{B}$ is subjected to a body force field $\mathbf{b}$. The boundary is partitioned into $\{\partial\mathcal{B}_\mathrm{u}, \partial\mathcal{B}_\mathrm{T}\}$ such that the displacement field $\mathbf{u}$ is prescribed on $\partial\mathcal{B}_\mathrm{u}$ and tractions $\mathbf{T}$ are prescribed on $\partial\mathcal{B}_{\mathrm{T}}$. Similarly, the boundary is partitioned into $\{\partial\mathcal{B}_\phi, \partial\mathcal{B}_{\mathrm{M}}\}$ such that the infinitesimal rotation vector field $\boldsymbol{\phi}$ is prescribed on $\partial\mathcal{B}_\phi$ and the couple traction $\mathbf{M}$ is prescribed on $\partial\mathcal{B}_{\mathrm{M}}$.

The kinematical vector fields $\mathbf{u}$ and $\boldsymbol{\phi}$ give rise to the micro-strain quantities ϵ_α, μ_α, and $\boldsymbol{\gamma}_\alpha$ corresponding to elongation, torsion and shear, respectively, of the doublet. The work conjugate micro-stresses p_α, m_α and $\mathbf{t}_\alpha$, corresponding to elongation, torsion and shear, respectively, are assumed to be derivable from a stored strain energy function $W = W(\epsilon_\alpha, \mu_\alpha, \gamma_{\alpha i})$ to exclude the possibility of generating energy through a closed cycle of deformation. Unless otherwise noted, Latin subscripts will denote the component of the quantity expressed with respect to a Cartesian coordinate system. For example, $\gamma_{\alpha i}$ are the Cartesian components of $\boldsymbol{\gamma}_\alpha$ with respect to the orthonormal basis $\mathbf{e}_i$. For the most general linear elastic response, the micro-stresses are related to the micro-strains through the linear constitutive relations (2.78), (2.79) and (2.80). For material stability we take the strain energy function W to be non-negative at all points for all compatible micro-strain fields. We shall return to this point below. Due to the resulting quadratic form, W achieves a minimum when $\epsilon_\alpha = 0$, $\mu_\alpha = 0$ and $\gamma_{\alpha i} = 0$; Without loss of generality we take this minimum to be zero. It follows that $W = 0$ if and only if $\epsilon_\alpha = \mu_\alpha = \gamma_{\alpha i} = 0$.

The internal energy W_{int} is given by (Granik and Ferrari 1993)

$$2\,W_{\mathrm{int}} = \sum_{\alpha=1}^{n} \int_{\mathcal{B}} (p_\alpha\, \epsilon_\alpha + m_\alpha\, \mu_\alpha + \mathbf{t}_\alpha \cdot \boldsymbol{\gamma}_\alpha)\, dV \tag{8.1}$$

and the external energy W_{ext} is given by

$$2\,W_{\mathrm{ext}} = \int_{\mathcal{B}} \mathbf{b}\cdot\mathbf{u}\,dV + \int_{\partial\mathcal{B}} (\mathbf{T}\cdot\mathbf{u} + \mathbf{M}\cdot\boldsymbol{\phi})\, dS. \tag{8.2}$$

At equilibrium, $W_{\mathrm{int}} = W_{\mathrm{ext}}$.

Let $\{\mathbf{u}^1, \boldsymbol{\phi}^1, \mu^1_\alpha, \gamma^1_\alpha, p^1_\alpha, m^1_\alpha, \mathbf{t}^1_\alpha\}$ and $\{\mathbf{u}^2, \boldsymbol{\phi}^2, \mu^2_\alpha, \gamma^2_\alpha, p^2_\alpha, m^2_\alpha, \mathbf{t}^2_\alpha\}$ denote two sets of fields which satisfy the governing equations. Furthermore, let $\tilde{(\cdot)}$ denote the difference in the quantity $(\cdot)$ between the two solutions. For example, $\tilde{\mathbf{u}} = \mathbf{u}^2 - \mathbf{u}^1$. It follows that

$$2\,\tilde{W}_{\mathrm{ext}} = \int_{\mathcal{B}} \tilde{\mathbf{b}}\cdot\tilde{\mathbf{u}}\,dV + \int_{\partial\mathcal{B}} \left(\tilde{\mathbf{T}}\cdot\tilde{\mathbf{u}} + \tilde{\mathbf{M}}\cdot\tilde{\boldsymbol{\phi}}\right) dS = 0. \tag{8.3}$$

Thus, at equilibrium $\tilde{W}_{\mathrm{int}} = 0$ from which it follows that $\tilde{\epsilon}_\alpha = \tilde{\mu}_\alpha = \tilde{\gamma}_{\alpha i} = 0$ which then implies that the micro-strains and thus also the micro-stresses are the same for the two solutions, i.e., $\epsilon^1_\alpha = \epsilon^2_\alpha, \ldots, \gamma^1_{\alpha i} = \gamma^2_{\alpha i}$. It follows that the two sets of kinematical fields differ by at most a motion which is strain-free. That is, the micro-strains associated with the kinematical fields $\tilde{\mathbf{u}}$ and $\tilde{\boldsymbol{\phi}}$ are zero. We call these micro-strain–free motions rigid body motions. The infinitesimal displacement $\mathbf{u}$ and rotation $\boldsymbol{\phi}$ fields are thus unique to within at most a rigid body motion. This concludes the uniqueness proof.

It was mentioned above that we demand the strain energy function W to be greater than or equal to zero when evaluated at any point within the body for all admissible motions. This was to assure material stability. In other words, W is to be non-negative for all physically realizable occurences of it arguements, the micro-strain.. In the realm of linear elastic CM, the strain energy functional W^{c} is taken to be a function of the linearized CM strain measure $\varepsilon_{ij}^{\mathrm{c}}$. To determine over what set W^{c} must be non-negative consider an $\varepsilon_{ij}^{\times} \in \mathbb{R}^6$. We now ask if there exists a displacement field u_i^{c} which when evaluated and *some* point yields the strain measure $\varepsilon_{ij}^{\times}$? The answer is in the affirmative; Take the displacement field $u_i^{\mathrm{c}} = \varepsilon_{ij}^{\times} x_j$ which yields a strain measure $\varepsilon_{ij}^{\mathrm{c}}$ at *all* points in the body. Thus, W^{c} must be non-negative for all $\varepsilon_{ij} \in \mathbb{R}^6$. The quadratic form of W^{c} in linear elasticity allows one to stipulate that $W^{\mathrm{c}} = 0$ if and only if $\varepsilon_{ij}^{\mathrm{c}} = 0$. As a result, W^{c} is required to be positive definite. In DM $(\epsilon_\alpha, \mu_\alpha, \gamma_{\alpha i}) \in \mathbb{R}^{8n}$ but it is not clear that it is necessary to demand that W be non-negative with respect to $\mathbb{R}^{8n}$—though this would be sufficient.

8.3 Inversion Technique

In this section we develop a connection between solutions in DM and CM. This connection proves useful in that it allows, given a solution in one realm, the generation of a solution in the other realm. Below we present conditions which are sufficient to allow for this connection.

Consider two mathematical models of the same physical body, one linear elastic DM and the other linear elastic CM. For the DM model we consider the non-scale theory with no infinitesimal rotational kinematical vector field ϕ. We assume the material to be incapable of supporting micro-torsional and micro-shear stresses. The equilibrium equations for the DM model are

$$\sum_{\alpha=1}^{n} \tau_{\alpha i}^{\circ} \tau_{\alpha j}^{\circ} p_{\alpha,j} + b_i = 0 \qquad \text{in } \mathcal{B} \tag{8.4}$$

where $p_{\alpha,j} := \partial p_\alpha / \partial x_j$ and $\tau_{\alpha i}^{\circ}$ is the direction cosine of the $\boldsymbol{\tau}_\alpha^{\circ}$ doublet with the x_i-axis. The boundary conditions are

$$n_j \sum_{\alpha=1}^{n} \tau_{\alpha i}^{\circ} \tau_{\alpha j}^{\circ} p_\alpha = \bar{T}_i \qquad \text{on } \partial\mathcal{B}_{\mathrm{T}} \tag{8.5}$$

and

$$u_i^d = \bar{u}_i \qquad \text{on } \partial\mathcal{B}_{\mathrm{u}} \tag{8.6}$$

where $\partial\mathcal{B}_{\mathrm{T}}$ and $\partial\mathcal{B}_{\mathrm{u}}$ form a partition of the boundary $\partial\mathcal{B}$ and $\bar{T}_i$ and $\bar{u}_i$ denote prescribed quantities. The micro-constitutive relation is

$$p_\alpha = \sum_{\beta=1}^{n} A_{\alpha\beta}\,\epsilon_\beta \tag{8.7}$$

where $A_{\alpha\beta}$ is symmetric and positive definite. We shall often make use of the nonpolar form of this constitutive relation, eqn. (5.3). We shall make use of the nonpolar constitutive relation (5.3). Compatibility is given by the identity

$$\varepsilon^{\rm d}_{11,22} - 2\,\varepsilon^{\rm d}_{12,12} + \varepsilon^{\rm d}_{22,11} = 0. \tag{8.8}$$

where $\varepsilon^{\rm d}_{ij} = (u^{\rm d}_{i,j} + u^{\rm d}_{j,i})/2$.

The equilibrium equations for the CM model are the familiar equations

$$\sigma^{\rm c}_{ij,j} + b_i = 0 \qquad \text{in } \mathcal{B}. \tag{8.9}$$

The boundary conditions are

$$\sigma^{\rm c}_{ij}\,n_j = \bar{T}_i. \qquad \text{on } \partial\mathcal{B}_{\rm T}. \tag{8.10}$$

and

$$u^{\rm c}_i = \bar{u}_i \qquad \text{on } \partial\mathcal{B}_{\rm u}. \tag{8.11}$$

The constitutive relation is

$$\sigma^{\rm c}_{ij} = C_{ijkl}\,\varepsilon^{\rm c}_{kl}, \tag{8.12}$$

where $C_{ijkl} = C_{klij} = C_{jikl}$ and $\mathbf{C}$ is positive definite.

The transition from micro- to macro-stresses is given by the relation

$$\sigma^{\rm d}_{ij} = \sum_{\alpha=1}^{n} \overset{\circ}{\tau}_{\alpha i}\,\overset{\circ}{\tau}_{\alpha j}\,p_\alpha. \tag{8.13}$$

As noted by Granik and Ferrari (1993) it is possible to define equivalent macro-stresses and -strains in terms of the micro-stresses and -strains, respectively. The micro- to macro-strain relation is given by

$$\epsilon_\alpha = \overset{\circ}{\tau}_{\alpha i}\,\overset{\circ}{\tau}_{\alpha j}\,u^{\rm d}_{i,j}. \tag{8.14}$$

The micro- to macro-stress relation (8.13) permits a convenient representation of the DM governing equations. Substituting eqn. (8.13) into the DM equilibrium equation (8.4) yields

$$\sigma^{\rm d}_{ij,j} + b_i = 0 \qquad \text{in } \mathcal{B} \tag{8.15}$$

provided that the doublet directions do not vary spatially. Substituting eqn. (8.13) into the traction boundary condition (8.5) yields the expression

$$\sigma^{\rm d}_{ij}\,n_j = \bar{T}_i. \qquad \text{on } \partial\mathcal{B}_{\rm T}. \tag{8.16}$$

The form equivalence between eqns. (8.15) and (8.9) and between eqns. (8.16) and (8.10) gives rise to the following result. If $\sigma^{\rm c}_{ij}$ is an admissible stress field then any set of micro-stresses p_α which yield $\sigma^{\rm d}_{ij} \equiv \sigma^{\rm c}_{ij}$ satisfy both the micro-stress equilibrium and the micro-stress traction boundary condition.

Conversely, if $\{p_\alpha\}$ is an admissible set of micro-stresses then $\sigma^{\mathrm{c}}_{ij} = \sigma^{\mathrm{d}}_{ij}$ is an admissible CM stress field.

Similarly, if $\varepsilon^{\mathrm{c}}_{ij}$ is an admissible strain field then micro-strains obtained from eqn. (8.14) using $\varepsilon^{\mathrm{d}}_{ij} = \varepsilon^{\mathrm{c}}_{ij}$ is an admissible micro-strain field. Conversely, if ϵ_α is an admissible micro-strain field consistent with the equivalent macroscopic strain field $\varepsilon^{\mathrm{d}}_{ij}$ then $\varepsilon^{\mathrm{c}}_{ij} = \varepsilon^{\mathrm{d}}_{ij}$ is an admissible CM strain field.

We now introduce some matrix notation which will allow for convenient representation of some of the above relations and later manipulations. Let $\hat{\boldsymbol{\epsilon}} := \{\epsilon_1, \epsilon_2, \ldots, \epsilon_n\}^T$ be the column vector of axial micro-strains and let $\hat{\mathbf{p}} := \{p_1, p_2, \ldots, p_n\}^T$ be the column vector of axial micro-stresses. The micro-constitutive relation (8.7) may thus be expressed as $\hat{\mathbf{p}} = \mathbf{A}\,\hat{\boldsymbol{\epsilon}}$.

We now restrict attention to planar problems. The developments below, however, are easily extended to three dimensions. In this context, let $\hat{\boldsymbol{\sigma}} := \{\sigma_{11}, \sigma_{22}, \sigma_{12}\}^T$ be the column vector of in-plane stresses. Likewise, let $\hat{\boldsymbol{\varepsilon}} := \{\varepsilon_{11}, \varepsilon_{22}, \varepsilon_{12} + \varepsilon_{21}\}^T$. The constitutive relation (8.12) takes the form $\hat{\boldsymbol{\sigma}} = \hat{\mathbf{C}}\,\hat{\boldsymbol{\varepsilon}}$ where $\hat{\mathbf{C}}$ is the appropriate matrix representation of $\mathbf{C}$ for the type of planar problem under consideration, i.e., plane strain or plane stress.

The micro-macro relation (8.13) may be expressed as

$$\hat{\boldsymbol{\sigma}}^d = \mathbf{M}\,\hat{\mathbf{p}}, \tag{8.17}$$

which implicitly defines the $3 \times n$ matrix $\mathbf{M}$. Similarly, the micro-macro strain relation (8.14) can be expressed as

$$\hat{\boldsymbol{\epsilon}} = \mathbf{M}^T\,\hat{\boldsymbol{\varepsilon}}^{\mathrm{d}}. \tag{8.18}$$

From the developments given above it follows directly that

$$\hat{\mathbf{p}} = \mathbf{M}^{-1}\,\hat{\boldsymbol{\sigma}}^{\mathrm{c}} \tag{8.19}$$

is an admissible micro-stress field provided that σ^{c}_{ij} is admissible and $\mathbf{M}$ is invertible. The matrix $\mathbf{M}$ is invertible if $n = 3$ and none of the three doublets are collinear. For $n = 3$, $\mathbf{M}$ is given by

$$\mathbf{M} = \begin{bmatrix} (\tau^{\circ}_{11})^2 & (\tau^{\circ}_{21})^2 & (\tau^{\circ}_{31})^2 \\ (\tau^{\circ}_{12})^2 & (\tau^{\circ}_{22})^2 & (\tau^{\circ}_{32})^2 \\ \tau^{\circ}_{11}\,\tau^{\circ}_{12} & \tau^{\circ}_{21}\,\tau^{\circ}_{22} & \tau^{\circ}_{31}\,\tau^{\circ}_{32} \end{bmatrix}. \tag{8.20}$$

Furthermore,

$$\hat{\boldsymbol{\epsilon}} = \mathbf{M}^T\,\hat{\boldsymbol{\varepsilon}}^{\mathrm{c}} \tag{8.21}$$

is an admissible micro-strain field provided that $\boldsymbol{\varepsilon}^{\mathrm{c}}_{ij}$ is admissible. Thus, if $\boldsymbol{\varepsilon}^{\mathrm{c}}_{ij}$ and $\boldsymbol{\sigma}^{\mathrm{c}}_{ij}$ are the equilibrium solution fields to the CM problem then eqns. (8.19) and (8.21) yield admissible micro-stress and micro-strain fields, respectively. In order for these micro-strain and micro-stress fields to be the solution to the DM problem they must be related through the constitutive relation (8.7) or equivalently,

$$\mathbf{M}\,\mathbf{A}\,\mathbf{M}^{\mathrm{T}} = \hat{\mathbf{C}}. \tag{8.22}$$

The proof follows by substituting eqns. (8.19) and (8.21) into $\hat{\mathbf{p}} = \mathbf{A}\,\hat{\varepsilon}$ which yields

$$\hat{\sigma}^c = \mathbf{M}\,\mathbf{A}\,\mathbf{M}^{\mathbf{T}}\,\hat{\varepsilon}^c \tag{8.23}$$

which is true if $\mathbf{M}\,\mathbf{A}\,\mathbf{M}^T = \hat{\mathbf{C}}$. If the micro-constitutive relation is non-polar then it can be proven that $\mathbf{M}\,\mathbf{A}\,\mathbf{M}^T = A_\circ\,\mathbf{M}\,\mathbf{M}^T$ is isotropic for all values of θ if and only if $\gamma = \pi/3$.

In summary, if we desire a DM solution to a planar problem with three doublets and $\gamma = \pi/3$ then the solution can be calculated directly from the associated CM problem with an isotropic material. Again, we emphasize that this method of inversion of the macro-stresses has been presented only for the planar case but it is applicable to 3-D problems as well. It should be noted, however, that there is no arrangement of 6 doublets in 3 dimensions that with a non-polar micro-constitutive relation yields $\mathbf{MAM}^T$ to be isotropic.

8.3.1 Homogeneous Deformations

In this section we present some homogeneous deformations of discrete materials. We begin by considering $n = 3$ with $\gamma = \pi/3$. Consider a plate of DM material modeled with three in-plane doublets with a structural angle of $\gamma = \pi/3$ (cf. Fig. 8.1). We now subject the material square to uniaxial tension (cf. Fig. 8.2) and pure shear (cf. Fig. 8.3). Using the method of macro-stress inversion the micro-stresses can be computed for arbitrary angles of θ—the inclination of the τ_3° doublet with the x_1-axis. The results are presented as follows. In Fig. 8.4 the micro-stresses are presented for uniaxial tension as a

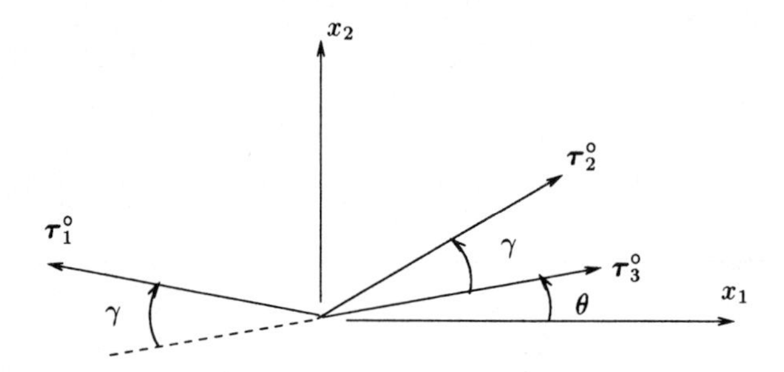

Fig. 8.1. Doublet geometry for homogeneous deformations

function of the angle θ. Note that compressive micro-stresses are achieved in distinction to the macroscopic principal stresses which are everywhere non-compressive. In Fig. 8.5 the micro-stresses for the pure shear loading case is presented. Note that the micro-stresses exceed the magnitude of the applied shear stress.

We now consider the effect of $\gamma \neq \pi/3$ for arbitrary θ. To simplify the presentation of results we will consider the energy stored in a plate of material

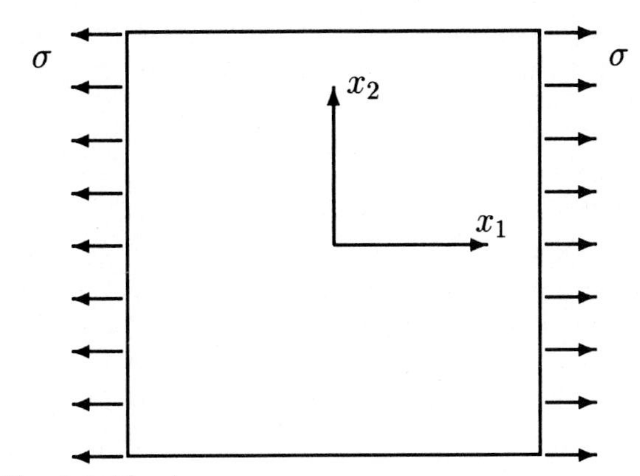

Fig. 8.2. Tension

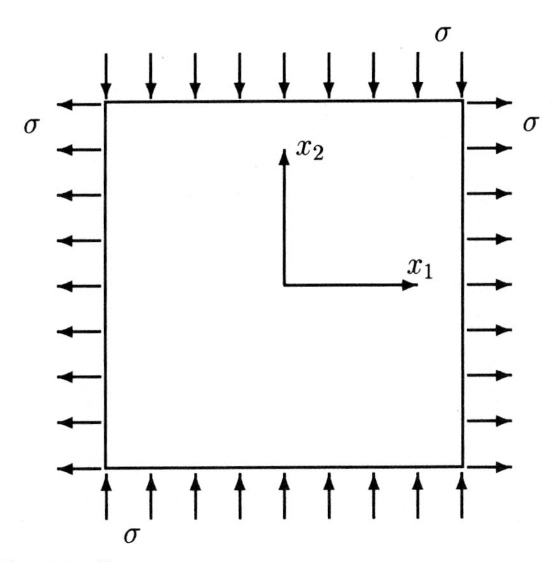

Fig. 8.3. Shear

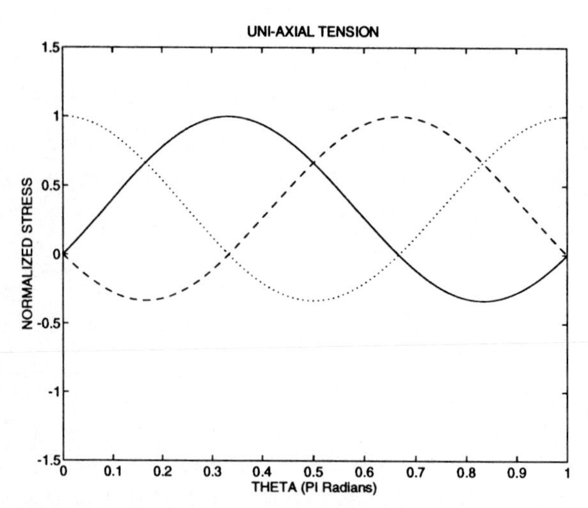

Fig. 8.4. Normalized micro-stress of a plate under uni-axial tension: solid-line is p_1/σ; dashed-line is p_2/σ; dotted-line is p_3/σ.

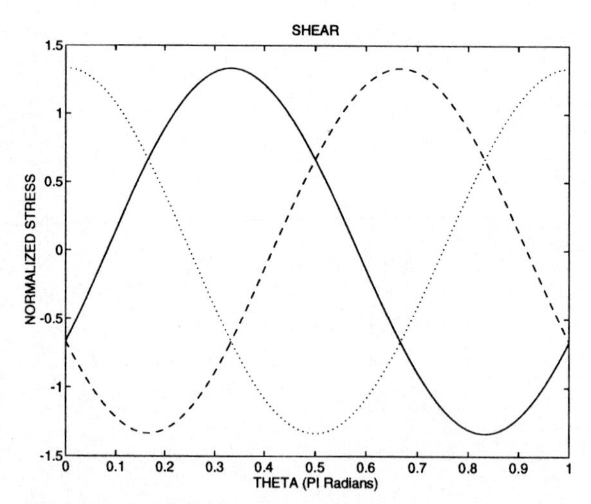

Fig. 8.5. Normalized micro-stress of a plate under shear: solid-line is p_1/σ; dashed-line is p_2/σ; dotted-line is p_3/σ.

subjected to shear. The stored energy is indicative of the magnitude of the stresses in the set of doublets. It is observed that γ has a significant effect on

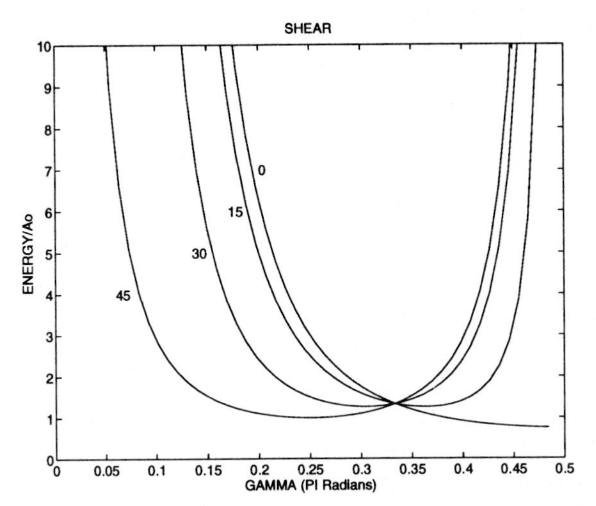

Fig. 8.6. Stored energy in a plate under shear for different values of θ (degrees).

the micro-stress field. In general, the energy grows unbounded as $\gamma \to 0$ and as $\gamma \to \pi/2$ indicating that at least one of the micro-stresses grows without bound. When $\gamma = \pi/3$ the internal stored energy is the same for all angles θ thus all curves illustrated in Fig. 8.6 pass through the point $(1/3, 4/3)$. For each value of θ there exists an angle γ which minimizes the internal stored energy. For a given value of θ the value of γ which minimizes the stored energy varies with the state of applied tractions.

8.3.2 Flamant Problem

Consider the classical problem of Flamant: A penetrating point force P acting normal to the straight boundary of a semi-infinite plate of isotropic material. The classical CM solution is characterized by a stress field whose principal stresses are everywhere non-positive. In this section we consider the DM solution to Flamant's problem with three doublets with a structural angle of $\gamma = \pi/3$ and for arbitrary θ. This problem has been treated previously by Granik and Ferrari (1993) for the case $\theta = 0$. Their solution contains some quantitative inconsistencies which are corrected here.

Since the DM domain consists of three doublets with $\gamma = \pi/3$ we can use the classical result of Flamant to obtain the DM solution using the method detailed in Sect. 8.3. Flamant's solution reads

$$\sigma^{\mathrm{c}}_{11} = -\frac{2P}{\pi}\frac{x_1^2 x_2}{(x_1^2 + x_2^2)^2} \tag{8.24}$$

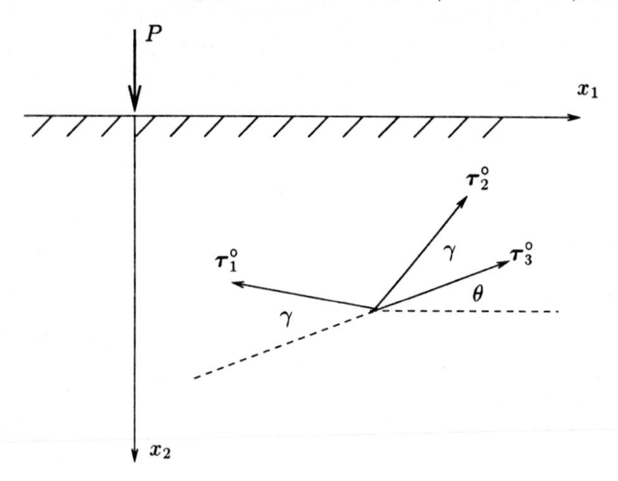

Fig. 8.7. Doublet geometry for Flamant's problem

$$\sigma_{12}^{c} = \sigma_{21}^{c} = -\frac{2P}{\pi}\frac{x_1\,x_2^2}{(x_1^2+x_2^2)^2} \tag{8.25}$$

$$\sigma_{22}^{c} = -\frac{2P}{\pi}\frac{x_2^3}{(x_1^2+x_2^2)^2} \tag{8.26}$$

and

$$u_1^{c} = \frac{P}{2\pi\mu}\left[\frac{\mu}{\lambda'+\mu}\left\{\arctan\left(\frac{x_2}{x_1}\right)-\frac{\pi}{2}\right\}+\frac{x_1\,x_2}{x_1^2+x_2^2}\right] \tag{8.27}$$

$$u_2^{c} = -\frac{P}{2\pi\mu}\left[\frac{\lambda'+2\mu}{2(\lambda'+\mu)}\log(x_1^2+x_2^2)+\frac{x_1^2}{x_1^2+x_2^2}\right]. \tag{8.28}$$

where $\lambda' := 2\lambda\mu/(\lambda+2\mu)$ and λ and μ are the Lamé constants of the isotropic continuum.

The micro-stresses for $\theta = 0$ are evaluated to be

$$p_1 = -\frac{4P}{3\pi}\frac{x_2^2(\sqrt{3}\,x_1+x_2)}{(x_1^2+x_2^2)^2} \tag{8.29}$$

$$p_2 = \frac{4P}{3\pi}\frac{x_2^2(\sqrt{3}\,x_1-x_2)}{(x_1^2+x_2^2)^2} \tag{8.30}$$

$$p_3 = -\frac{2P}{3\pi}\frac{x_2(3x_1^2-x_2^2)}{(x_1^2+x_2^2)^2}. \tag{8.31}$$

The classical solution due to Flamant is characterized by a stress field for which the principal stresses are nowhere tensile for a penetrating applied load. Unlike the Flamant stresses, the micro-stresses are found to be tensile within particular regions of the domain. For instance, the p_3 micro-stress associated with the $\boldsymbol{\tau}_3$ doublet is tensile within the sector defined by $x_2 > \sqrt{3}\,|x_1|$. The p_1 and p_2 micro-stresses are also tensile within specific regions.

To graphically illustrate these characteristics it is convenient to normalize eqns. (8.29)–(8.31). Let $\bar{x} := x_1/x_2$. It follows that eqns. (8.29)–(8.31) can be recast in the following form:

$$P1 := \frac{3\pi\, x_2}{4P} p_1 = -\frac{\sqrt{3}\,\bar{x}+1}{(1+\bar{x}^2)^2} \tag{8.32}$$

$$P2 := \frac{3\pi\, x_2}{4P} p_2 = \frac{\sqrt{3}\,\bar{x}-1}{(1+\bar{x}^2)^2} \tag{8.33}$$

$$P3 := \frac{3\pi\, x_2}{4P} p_3 = \frac{1-3\,\bar{x}^2}{2(1+\bar{x}^2)^2} P3. \tag{8.34}$$

In addition, the only non-zero principal macro-stress σ^{P} can be expressed as

$$S := \frac{3\pi\, x_2}{4P}\sigma^{\mathrm{P}} = -\frac{3}{2(1+\bar{x}^2)}. \tag{8.35}$$

The normalized quantities $P1$, $P2$, $P3$ and S are presented in Fig. 8.8 for $\theta = 0$.

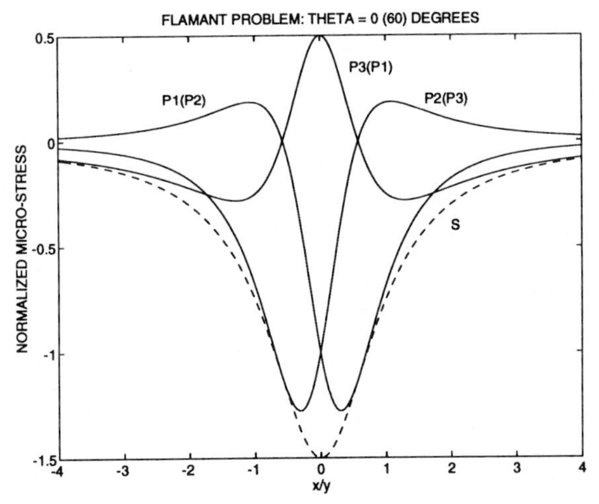

Fig. 8.8. Normalized micro- and macro-stresses for the Flamant problem with $\theta = 0$ ($\theta = \pi/3$).

We now provide the in-plane displacement field. Using the non-polar constitutive relation it may be verified that the components of the in-plane displacement field are given by

$$u_1^{\mathrm{d}} = \frac{2P}{3\pi A_{\circ}}\left[\arctan\left(\frac{x_2}{x_1}\right) - \frac{\pi}{2} + \frac{2\,x_1\,x_2}{x_1^2+x_2^2}\right] \tag{8.36}$$

$$u_2^{\mathrm{d}} = -\frac{2P}{3\pi A_{\circ}}\left[\frac{3}{2}\log(x_1^2+x_2^2) + \frac{2\,x_1^2}{x_1^2+x_2^2}\right]. \tag{8.37}$$

In eqn. (8.37), log denotes the natural logarithm. We note that the displacement field does not tend to zero at infinite distances from the point of application of the load as is the case in the Flamant solution (see e.g., Love (1944, p.211)). In general, the displacement field given by eqns. (8.36) and (8.37) is not the same displacement field as the Flamant solution. This observation is intuitive since the material domain of Flamant is characterized by two constitutive parameters while the microstructured material considered herein is characterized by only one constitutive parameter, namely $A_\circ$, while the stress field is independent of the constitutive relation. Let (λ, μ) denote the Lamé constants of the elastic continuum utilized be Flamant. When $(\lambda, \mu) \mapsto (3A_\circ/4, 3A_\circ/8)$ (i.e., Poisson's ratio is 1/3) it may be shown that the two displacement fields are equivalent.

To illustrate the effect of θ on the micro-stresses, plots similar to that in Fig. 8.8 are given for $\theta = 15, 30, 45$, degrees, in Figs. 8.9, 8.10 and 8.11, respectively. Closed form expressions for the micro-stresses in terms of θ

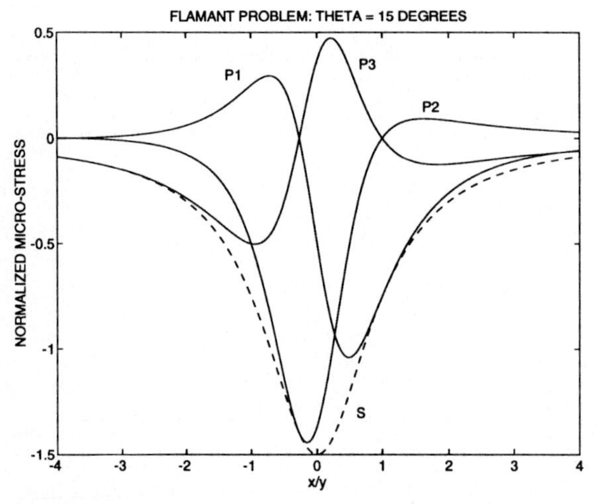

Fig. 8.9. Normalized micro- and macro-stresses for the Flamant problem with $\theta = 15$ degrees.

are very lengthy but for special values of θ simple expressions do exist. We have seen this above for $\theta = 0$. When $\theta = \pi/6$ similar expressions exist. The normalized micro-stresses when $\theta = \pi/6$ are given by

$$P1 := \frac{3\pi\, x_2}{4P} p_1 = -\frac{\bar{x}(\sqrt{3} + \bar{x})}{(1 + \bar{x}^2)^2} \tag{8.38}$$

$$P2 := \frac{3\pi\, x_2}{4P} p_2 = \frac{\bar{x}^2 - 3}{2(1 + \bar{x}^2)^2} \tag{8.39}$$

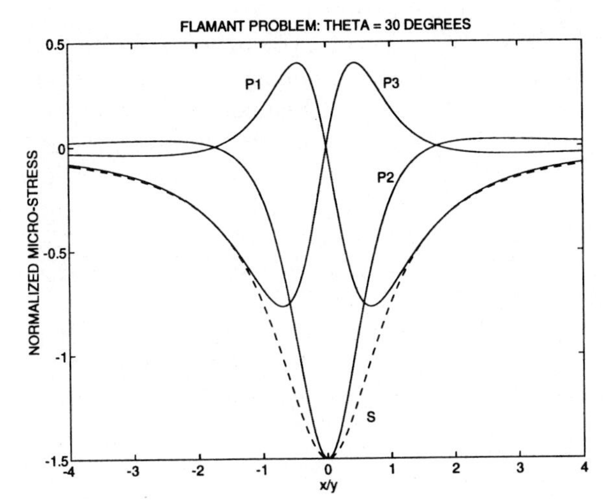

Fig. 8.10. Normalized micro- and macro-stresses for the Flamant problem with $\theta = 30$ degrees.

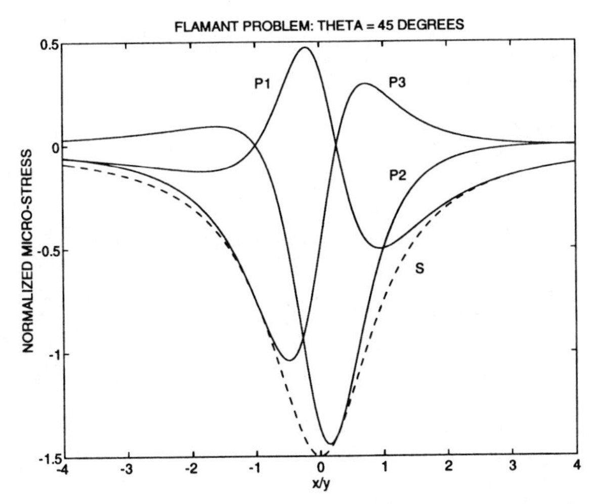

Fig. 8.11. Normalized micro- and macro-stresses for the Flamant problem with $\theta = 45$ degrees.

$$P3 := \frac{3\pi\, x_2}{4P} p_3 = \frac{\bar{x}(\sqrt{3}-\bar{x})}{(1+\bar{x}^2)^2}. \tag{8.40}$$

These are the functions plotted in Fig. 8.10.

We note that the components of the displacement field for the DM solutions for all values of θ are given by eqns. (8.36) and (8.37).

8.3.3 Kelvin's Problem

The problem of Flamant treated in Sect. 8.3.2 is a Green's function for elasticity problems with normal surface tractions. We now present another Green's function: a point force acting in the plane of an infinite plate, originally solved by Lord Kelvin. This solution enables one to solve problems with arbitrary body force distributions by integration. In particular, let us apply a point force of magnitude P at the origin of the plate in the negative x_1 direction, with the x_1- and x_2-axes lying in the plane of the plate (cf. Fig. 8.12). From

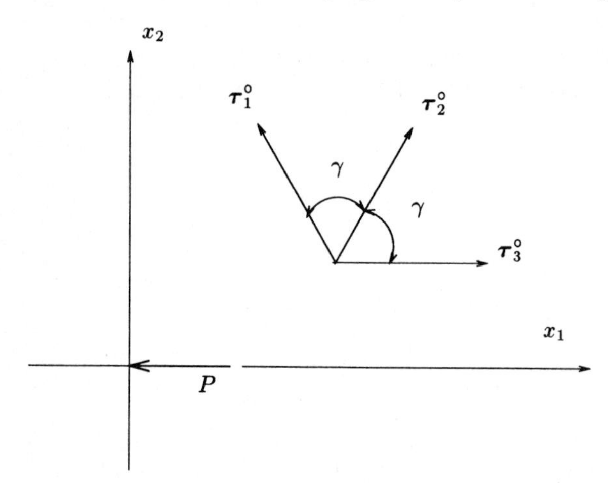

Fig. 8.12. Doublet geometry for Kelvin's problem

classical CM mechanics, the macrostress state is given by (Love 1944)

$$\sigma_{11}^c = \frac{P}{8\pi}\frac{x_1}{r^2}\left[7-6\left(\frac{x_2}{r}\right)^2\right] \tag{8.41}$$

$$\sigma_{22}^c = \frac{P}{8\pi}\frac{x_1}{r^2}\left[6\left(\frac{x_2}{r}\right)^2-1\right] \tag{8.42}$$

$$\sigma_{12}^c = \frac{P}{8\pi}\frac{x_2}{r^2}\left[6\left(\frac{x_1}{r}\right)^2+1\right] \tag{8.43}$$

where $r^2 := x_1^2 + x_2^2$.

The micro-stresses are obtained by substituting the CM macro-stresses (8.41)–(8.43) into eqn. (8.19) yielding

$$p_1 = -\frac{P}{12\pi r^4}\left[x_1^3 - 5\,x_1\,x_2^2 + 7\sqrt{3}\,x_1^2\,x_2 + \sqrt{3}\,x_2^3\right] \tag{8.44}$$

$$p_2 = -\frac{P}{12\pi\,r^4}\left[x_1^3 - 5\,x_1\,x_2^2 - 7\sqrt{3}\,x_1^2\,x_2 - \sqrt{3}\,x_2^3\right] \tag{8.45}$$

$$p_3 = \frac{P\,x_1}{12\pi\,r^4}\left[11\,x_1^2 - x_2^2\right] \tag{8.46}$$

where we have assumed a three doublet DM lattice with $\gamma = \pi/3$ and $\theta = 0$. Expectedly, eqns. (8.44)–(8.46) satisfy the equilibrium eqns. (8.4), and the microstrains derived via eqn. (5.3) satisfy the equation of compatibility (8.8).

8.3.4 Stress Concentration

In addition to obtaining DM Green's functions, the Inversion technique can be used to find microstress concentration factors. As an example, we consider a circular void within the infinite plate subjected to hydrostatic tractions, $\mathbf{T} = \sigma\,\mathbf{n}$, at infinity, as illustrated in Fig. 8.13. The CM solution in polar

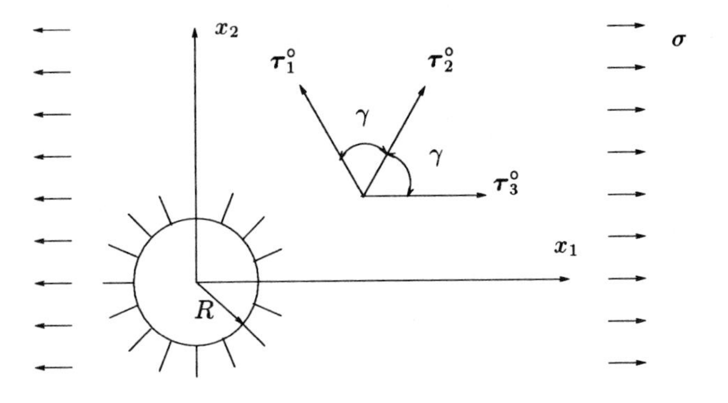

Fig. 8.13. Circular hole in an infinite plate.

coordinates is given by

$$\sigma_{rr}^c = \sigma\left[1 - (R/r)^2\right] \tag{8.47}$$

$$\sigma_{\theta\theta}^c = \sigma\left[1 + (R/r)^2\right] \tag{8.48}$$

$$\sigma_{r\theta}^c = 0 \tag{8.49}$$

where r and θ are polar coordinates and the circular void is centered at $r = 0$ with radius R.

To obtain the microstresses from eqns. (8.47)–(8.49) we must first rewrite the macrostresses in terms of Cartesian coordinates, since the DM lattice is defined with respect to this latter coordinate system. Equation (8.19) can then be utilized to give the microstresses around the void

$$p_1 = \frac{2\sigma}{3}\left[1 - \frac{R^2\,(x_2^2 - 2\sqrt{3}\,x_1\,x_2 - x_1^2)}{r^4}\right] \tag{8.50}$$

$$p_2 = \frac{2\sigma}{3}\left[1 - \frac{R^2\,(x_2^2 + 2\sqrt{3}\,x_1\,x_2 - x_1^2)}{r^4}\right] \tag{8.51}$$

$$p_3 = \frac{2\sigma}{3}\left[1 - \frac{2\,R^2\,(x_1^2 - x_2^2)}{r^4}\right]. \tag{8.52}$$

Comparison of the macro- and microstresses shows that while CM predicts solely compressive principal stresses arising from applied compressive tractions, the microstresses are tensile in certain regions adjacent to the void. It should thus not be surprising if a granular body under hydrostatic pressure develops tensile openings, analogous to the paradox associated with Flamant's Problem. We find that the microstresses in the vicinity of the void vary from -1 to 3 times the far-field microstresses while the macrostresses vary between 0 and 2 times the far-field equivalent macro-stress.

8.4 Micro-stress Function

Consider a non-polar medium with no body forces and with lattice geometry as shown in, Fig. 8.1a. The equilibrium eqns. (8.4) take the form

$$(p_1 + p_2 + \csc^2\gamma\, p_3)_{,1} + \tan\gamma\,(p_2 - p_1)_{,2} = 0 \tag{8.53}$$

$$(p_2 - p_1)_{,1} + \tan\gamma\,(p_1 + p_2)_{,2} = 0. \tag{8.54}$$

The Integrability theorem applied to eqn. (8.53) implies the existence of a function $\Psi = \Psi(x_1, x_2)$ such that

$$\Psi_{,1} = \tan\gamma\,(p_2 - p_1) \tag{8.55}$$

$$\Psi_{,2} = -(p_1 + p_2 + \csc^2\gamma\, p_3)_{,1} + \tan\gamma\,(p_2 - p_1). \tag{8.56}$$

Similarly, the Integrability theorem applied to eqn. (8.54) implies the existence of a function $\Theta = \Theta(x_1, x_2)$ such that

$$\Theta_{,1} = -\tan^2\gamma\,(p_1 + p_2) \tag{8.57}$$

$$\Theta_{,2} = \tan\gamma\,(p_2 - p_1). \tag{8.58}$$

From eqns. (8.55) and (8.58) follows the relation $\Psi_{,1} = \Theta_{,2}$ which itself implies the existence of a function $\chi = \chi(x_1, x_2)$ such that $\chi_{,1} = \Theta$ and $\chi_{,2} = \Psi$. Solving for the micro-stresses in terms of the second partial derivatives of χ yields

$$p_1 = -\frac{1}{2}\cot^2\gamma\,(\tan\gamma\,\chi_{,12} + \chi_{,11}) \tag{8.59}$$

$$p_2 = -\frac{1}{2}\cot^2\gamma\,(\tan\gamma\,\chi_{,12} - \chi_{,11}) \tag{8.60}$$

$$p_3 = \cot^2\gamma\,\left(\cos^2\gamma\,\chi_{,11} - \sin^2\gamma\,\chi_{,22}\right) \tag{8.61}$$

The micro-stresses (8.59)–(8.61) thus satisfy equilibrium for a sufficiently differentiable function χ. Compatibility is now addressed. Compatibility in terms of the micro-stresses is obtained by substituting the constitutive relation $p_\alpha = A_\circ\,\epsilon_\alpha$ into eqn. (8.8):

$$0 = \left[\sec^2\gamma\,(p_1+p_2) - 2\cot^2\gamma\,p_3\right]_{,11} + 2\,p_{3,22} + \csc\gamma\,\sec\gamma\,(p_1-p_2)_{,12}. \tag{8.62}$$

Substitution of the micro-stresses (8.59)–(8.61) into the compatibility relation (8.62) yields

$$0 = \cot^2\gamma\,\csc^2\gamma\,(1+\cos^4\gamma)\chi_{,1111} + \csc^2\gamma(1-4\cos^4\gamma)\chi_{,1122} + 2\cos^2\gamma\,\chi_{,2222} \tag{8.63}$$

Any function χ which satisfies eqn. (8.63) thus yields a micro-stress field which satisfies equilibrium and compatibility. When $\gamma = \pi/3$, eqn. (8.63) simplifies to

$$0 = \chi_{,1111} + 2\,\chi_{,1122} + \chi_{,2222} =: \nabla^2\cdot\nabla^2\chi =: \nabla^4\chi \tag{8.64}$$

which is the bi-harmonic equation.

Substituting the micro-stresses (8.59)–(8.61) into the micro- to macro-stress relation (8.13) yields

$$\sigma_{11}^d = \bar{\chi}_{,22} \tag{8.65}$$

$$\sigma_{22}^d = \bar{\chi}_{,11} \tag{8.66}$$

$$\sigma_{12}^d = -\bar{\chi}_{,12} \tag{8.67}$$

where $\bar{\chi} := -\cos^2\gamma\,\chi$. Equations (8.65)–(8.67) are form equivalent to the equations for the classical continuum stresses in terms of the Airy stress function. When $\gamma = \pi/3$ the connection between DM and CM is further elucidated, since χ is governed by the bi-harmonic equation, as the stresses are obtained in CM from equations which are form equivalent to eqns. (8.65)–(8.67).

In Sects. 8.4.1 through 8.4.4 below, we investigate the solutions generated by the family of third-order polynomial $\chi(x_1, x_2)$ functions, and we present the MSF for the problems solved above via the inversion technique of Sect. 8.3.

8.4.1 Third-Order Polynomial χ-Functions

Considering the combined equilibrium/compatibility relation (8.63) shows that any third-order polynomial $\chi(x_1, x_2)$ represents an admissible DM solution. Let us consider the DM lattice with $\gamma = \pi/3$ and postulate a MSF of the form:

$$\chi(x_1, x_2) = A_1\,x_1 + A_2\,x_1^2 + A_3\,x_1^3 + B_1\,x_2 + B_2\,x_2^2 + B_3\,x_2^3 + C_1\,x_1\,x_2 + C_2\,x_1^2\,x_2 + C_3\,x_1\,x_2^2 + D. \tag{8.68}$$

We substitute eqn. (8.68) into eqns. (8.59)–8.61 to obtain the resulting microstresses:

$$p_1 = -\frac{1}{3}\left[\left(\sqrt{3}\,C_2 + 3\,A_3\right)\,x_1 + \left(\sqrt{3}\,C_3 + C_2\right)\,x_2 + \frac{\sqrt{3}}{2}\,C_1 + A_2\right] \tag{8.69}$$

$$p_2 = \frac{1}{3}\left[\left(\sqrt{3}\,C_2 - 3\,A_3\right)\,x_1 + \left(\sqrt{3}\,C_3 - C_2\right)\,x_2 + \frac{\sqrt{3}}{2}\,C_1 - A_2\right] \tag{8.70}$$

$$p_3 = \frac{1}{6}\left[3\,(A_3 - C_3)\,x_1 + (C_2 - 9\,B_3)\,x_2 - 3\,B_2 + A_2\right]. \tag{8.71}$$

Substituting eqns. (8.69)–(8.71) into the relation (8.13) yields the equivalent macrostresses:

$$\sigma_{11}^d = -\frac{1}{2}\left[C_3\,x_1 + 3\,B_3\,x_2 + B_2\right] \tag{8.72}$$

$$\sigma_{22}^d = -\frac{1}{2}\left[3\,A_3\,x_1 + C_2\,x_2 + A_2\right] \tag{8.73}$$

$$\sigma_{12}^d = \frac{1}{4}\left[2\,C_2\,x_1 + 2\,C_3\,x_2 + C_1\right]. \tag{8.74}$$

Analysing either the micro- or macrostresses, we see that χ given by eqn. (8.68) will provide solutions to problems with homogeneous or linearly varying stress states. Examples include uniaxial or biaxial tension and compression, pure shear and beams under pure bending. Note that shear stresses arise only from the seventh through the ninth terms of eqn. (8.68). Hence, if the three C_i coefficients are zero, the x_1- and x_2-axes are the pincipal axes; and conversely, states of shear are given by only the non-zero C_i's. The coefficients A_1, B_1 and D have no effect on the stress state and thus represent superfluous information.

8.4.2 Flamant's Problem

The family of third-order polynomials given by eqn. (8.68) represents only a subset of admissible χ functions. We now consider a more involved MSF containing a trigonometric term:

$$\chi(x_1, x_2) = \frac{4\,P\,x_1}{\pi}\left[1 - \arctan(x_2/x_1)\right] \tag{8.75}$$

which represents an admissible DM solution when $\gamma = \pi/3$. The microstresses derived from eqn. (8.75) are identical with eqns. (8.29)–(8.31), indicating that we have found the MSF for Flamant's Problem. Note, however, that for this particular application of the MSF a slightly varied form of eqns. (8.59)–(8.61) is necessary, since the coordinate system has been changed to that shown in Fig. 8.7.

8.4.3 Kelvin's Problem

The point force in the infinite plane is a similar problem in some respects to Flamant's Problem. Hence, we start our search for the appropriate MSF with modifications to the family of arc tangent functions. The final result is

$$\chi(x_1, x_2) = \frac{P}{4\pi}\left[x_1 \log(x_1^2 + x_2^2) + 8\,x_2 \arctan(x_1/x_2) - 8\,x_1\right]. \tag{8.76}$$

Substituting eqn. (8.76) into eqns. (8.59)–(8.61) yields microstresses identical to eqns. (8.44)–(8.46).

8.4.4 Stress Concentration

The problem of a circular void within the hydrostatically stressed plate is the superposition of a homogeneous stress state with the concentration stresses arising around the hole. Thus, the MSF should be a sum of a second-order polynomial and a term or terms that account for the effects of the void. The result we find is

$$\chi(x_1, x_2) = -2\,\sigma\left[x_1^2 + x_2^2 - R^2 \log\left(x_1^2 + x_2^2\right)\right] \tag{8.77}$$

which exactly recovers eqns. (8.50)–(8.52). Expectedly, the logarithmic term acts to satisfy the zero-traction boundary condition at surface of the void.

8.5 Closure

We initiated our study by presenting a uniqueness theorem in linear elastic DM. We then developed two methods for obtaining solutions in isotropic, plane elastostatics.

In the first, we noted a correspondence between DM and CM for specific DM lattice geometries, namely, a three doublet arrangement with $\gamma = \pi/3$. The result of this connection between DM and CM is that given a solution in either one of the two regimes, one can generate an equivalent solution in the other. We demonstrated the utility of this technique by obtaining DM solutions to homogeneous deformation problems, the classic problems of Flamant and Kelvin, and stress concentration around a hole. In the first two applications, we extended the study to a lattice rotated by an angle θ with respect to the original coordinate system. In the case of homogeneous deformations, we also analyzed the general three double lattice where $\gamma \neq \pi/3$.

The second technique which was developed arises when the micro-stress equilibrium and micro-strain compatibility requirements are manipulated to yield the microstress function (MSF). In illustrating the use of this second technique, we studied the solutions generated by the family of third order polynomial MSF's, and we derived the MSF's corresponding to the three problems considered via the inversion technique.

9. Multi-scale Solutions

M. Ferrari and A. Imam

9.1 Introduction

The governing field equations in doublet mechanics involve scaling parameters η_α which, for every doublet α, represent the distance between two adjacent nodes in the underlying Bravais lattice comprising the body. Here, $\alpha = 1, \ldots, n$, where n is the valence of the Bravais lattice. We expect the physical response of bodies modeled by doublet mechanics, such as particulate and granular media, under various types of loading to depend on the scaling parameters η_α. Therefore, it is of interest to obtain solutions in doublet mechanics that exhibit dependence on scaling parameters.

In some of the previous chapters, particularly in Chap. 8, solutions to various boundary value problems were presented in which the governing equations were based on the first level of approximation in doublet mechanics. At such a level of approximation, the scaling parameters do not appear and, consequently, the solutions are scale-independent. Indeed, the governing field equations at the first level of approximation for non-polar media correspond to those of continuum elasticity in which it is well known that no scaling parameters are present.

At higher levels of approximation, the doublet-mechanical solutions generally involve the scaling parameters. In this chapter we present a formulation whereby doublet-mechanical solutions for the second level of approximation can be obtained. In Sect. 9.2 we recall the general field equations and boundary conditions from Chap. 1. Then by expanding the displacement field in a convergent power series, we develop governing equations along with appropriate boundary conditions whose solutions represent the displacement field at the second level of approximation. These equations indicate that it is possible for a boundary value problem at the second level of approximation to have a solution which is independent of the scaling parameters.

The expansion of the displacement field gives rise to a series of equlibrium boundary value problems whose well-posedness is examined in Sect. 9.3. In particular, we show that the necessary conditions for existence of solutions for such equilibrium boundary value problems are satisfied. Finally, in Sect. 9.4 we solve a particular boundary value problem in order to illustrate the application of the present formulation in obtaining multi-scale solutions.

It is also possible to treat multi-scale problems within the context of generalized theories of continuum mechanics. A comparative study of such theories and doublet mechanics is presented in Chap. 4 of this monograph.

9.2 Field Equations and Boundary Conditions

We represent a body B by an array of nodes whose interactions are purely central so that the torsional and shear microstresses vanish everywhere. Furthermore, we assume the interactions are local and homogeneous. The former implies that the elongation microstress p_α, in a typical doublet (A, B_α), depends only on the elongation microstrain ϵ_α and is independent of the elongation microstrains in the other doublets emanating from the same node A. For homogeneous interactions, there is only one micromodulus $A_\circ$ and, as a consequence, the constitutive relation is expressed by

$$p_{\alpha i} = A_\circ \, \epsilon_{\alpha i}, \tag{9.1}$$

as elaborated in Sect. 5.2.

The linearized field equations of doublet mechanics are given by eqn. (1.33). This equation was expressed, after some manipulation, as eqn. (5.13) which is subsequently used in this chapter. For the equilibrium problem, this equation is rewritten as

$$A_\circ \sum_{\alpha=1}^{n} \overset{\circ}{\tau}_{\alpha i} \overset{\circ}{\tau}_{\alpha j} \sum_{\delta=2,4,\ldots}^{R} \frac{(\eta_\alpha)^{\delta-2}}{\delta!} \overset{\circ}{\tau}_{\alpha k_1} \cdots \overset{\circ}{\tau}_{\alpha k_\delta} \frac{\partial^\delta u_j}{\partial x_{k_1} \ldots \partial x_{k_\delta}} = 0, \tag{9.2}$$

where summation convention is implied only for repeated Latin indices.

In what follows, we restrict the analysis to the lowest level of approximation containing the scaling parameters η_α. In eqn. (9.2), this corresponds to $R = 4$ which we will refer to as the second level of approximation. As a result, the expanded version of eqn. (9.2) can be written as

$$A_\circ \sum_{\alpha=1}^{n} \overset{\circ}{\tau}_{\alpha i} \overset{\circ}{\tau}_{\alpha j} \overset{\circ}{\tau}_{\alpha k} \overset{\circ}{\tau}_{\alpha l} \left(\frac{1}{2!} u_{j,kl} + \frac{\eta_\alpha^2}{4!} \overset{\circ}{\tau}_{\alpha p} \overset{\circ}{\tau}_{\alpha q} u_{j,klpq} \right) = 0, \tag{9.3}$$

where comma signifies partial differentiation.

Next, the constitutive equation in eqn. (9.1) is written in terms of the displacement field. To this end, we note that the microstrain ϵ_α is given as a function of the displacement in eqn. (1.12), wherein the second level of approximation corresponds to $M = 2$. Using this equation, eqn. (9.1) is rewritten as

$$p_{\alpha i} = A_\circ \, \epsilon_\alpha \, \overset{\circ}{\tau}_{\alpha i} = A_\circ \, \overset{\circ}{\tau}_{\alpha i} \overset{\circ}{\tau}_{\alpha j} \overset{\circ}{\tau}_{\alpha k_1} \left(u_{j,k_1} + \frac{\eta_\alpha}{2!} \overset{\circ}{\tau}_{\alpha k_2} u_{j,k_1 k_2} \right). \tag{9.4}$$

We consider a boundary value problem where the traction vector $\mathbf{T}$ is prescribed everywhere on the boundary of the body, denoted by ∂B. The appropriate form of the traction boundary conditions was developed in eqns. (1.35),

(1.37) and (1.38). Here, since $M = 2$, r has a value equal to 1 and, consequently, eqn. (1.37) implies

$$n_{k_1} \sum_{\alpha=1}^{n} \overset{\circ}{\tau}_{\alpha k_1} \left(p_{\alpha i} - \frac{\eta_\alpha}{2!} \overset{\circ}{\tau}_{\alpha k_2} \frac{\partial p_{\alpha i}}{\partial x_{k_2}} \right) = T_i. \tag{9.5}$$

Substituting $p_{\alpha i}$ from eqn. (9.4) into eqn. (9.5) and making some simplification results in

$$A_\circ \, n_l \sum_{\alpha=1}^{n} \overset{\circ}{\tau}_{\alpha l} \, \overset{\circ}{\tau}_{\alpha i} \, \overset{\circ}{\tau}_{\alpha j} \, \overset{\circ}{\tau}_{\alpha k_1} \left(u_{j,k_1} - \frac{\eta_\alpha^2}{4} \overset{\circ}{\tau}_{\alpha k_2} \, \overset{\circ}{\tau}_{\alpha k_3} \, u_{j,k_1 k_2 k_3} \right) = T_i. \tag{9.6}$$

We also assume that the separation between adjacent nodes of a doublet is the same for all the doublets originating from the same node. That is,

$$\eta_\alpha = \eta, \qquad \alpha = 1, 2, \ldots, n \tag{9.7}$$

which results in a single scaling parameter η.

Now, let the applied traction vector have a convergent power series expansion, in the small scaling parameter η, whose first three terms are given by

$$\mathbf{T} = \mathbf{T}^{(0)} + \eta \, \mathbf{T}^{(1)} + \eta^2 \, \mathbf{T}^{(2)}. \tag{9.8}$$

Then, we assume that the displacement field of the above traction boundary value problem admits a convergent power series expansion in η, whose first few terms are given by

$$\mathbf{u} = \mathbf{v} + \eta \, \mathbf{z} + \eta^2 \, \mathbf{w}. \tag{9.9}$$

Substituting eqn. (9.9) into eqn. (9.3) results in

$$\begin{aligned} A_\circ \sum_{\alpha=1}^{n} \overset{\circ}{\tau}_{\alpha i} \, \overset{\circ}{\tau}_{\alpha j} \, \overset{\circ}{\tau}_{\alpha k} \, \overset{\circ}{\tau}_{\alpha l} \Bigg[& \frac{1}{2!} \left(v_{j,kl} + \eta \, z_{j,kl} + \eta^2 \, w_{j,kl} \right) \\ & + \frac{\eta^2}{4!} \overset{\circ}{\tau}_{\alpha p} \, \overset{\circ}{\tau}_{\alpha q} \left(v_{j,klpq} + \eta \, z_{j,klpq} + \eta^2 \, w_{j,klpq} \right) \Bigg] = 0, \end{aligned} \tag{9.10}$$

from which three equations, corresponding to the first three powers of η, are found to be

$$A_\circ \sum_{\alpha=1}^{n} \overset{\circ}{\tau}_{\alpha i} \, \overset{\circ}{\tau}_{\alpha j} \, \overset{\circ}{\tau}_{\alpha k} \, \overset{\circ}{\tau}_{\alpha l} \, v_{j,kl} = 0, \tag{9.11}$$

$$A_\circ \sum_{\alpha=1}^{n} \overset{\circ}{\tau}_{\alpha i} \, \overset{\circ}{\tau}_{\alpha j} \, \overset{\circ}{\tau}_{\alpha k} \, \overset{\circ}{\tau}_{\alpha l} \, z_{j,kl} = 0, \tag{9.12}$$

$$A_\circ \sum_{\alpha=1}^{n} \overset{\circ}{\tau}_{\alpha i} \, \overset{\circ}{\tau}_{\alpha j} \, \overset{\circ}{\tau}_{\alpha k} \, \overset{\circ}{\tau}_{\alpha l} \left(w_{j,kl} + \frac{1}{12} \overset{\circ}{\tau}_{\alpha p} \, \overset{\circ}{\tau}_{\alpha q} \, v_{j,klpq} \right) = 0. \tag{9.13}$$

Similarly, when eqns. (9.8) and (9.9) are substituted into eqn. (9.6), three equations corresponding to the first three powers of η are found to hold on the boundary of the body. They are given as

$$A_\circ n_l \sum_{\alpha=1}^{n} \overset{\circ}{\tau}_{\alpha l} \overset{\circ}{\tau}_{\alpha i} \overset{\circ}{\tau}_{\alpha j} \overset{\circ}{\tau}_{\alpha k} v_{j,k} = T_i^{(0)}, \tag{9.14}$$

$$A_\circ n_l \sum_{\alpha=1}^{n} \overset{\circ}{\tau}_{\alpha l} \overset{\circ}{\tau}_{\alpha i} \overset{\circ}{\tau}_{\alpha j} \overset{\circ}{\tau}_{\alpha k} z_{j,k} = T_i^{(1)}, \tag{9.15}$$

$$A_\circ n_l \sum_{\alpha=1}^{n} \overset{\circ}{\tau}_{\alpha l} \overset{\circ}{\tau}_{\alpha i} \overset{\circ}{\tau}_{\alpha j} \overset{\circ}{\tau}_{\alpha k} \left(w_{j,k} - \frac{1}{4} \overset{\circ}{\tau}_{\alpha p} \overset{\circ}{\tau}_{\alpha q} v_{j,kpq} \right) = T_i^{(2)}. \tag{9.16}$$

Expressions (9.11)–(9.13) represent three field equations for the displacements $\mathbf{v}$, $\mathbf{z}$ and $\mathbf{w}$, respectively, subject to the traction boundary conditions given in eqns. (9.14)–(9.16). Once these are solved, the actual displacement field $\mathbf{u}$ is determined, using eqn. (9.9), to within the accuracy of terms of order η^3. However, usually, the tractions prescribed at the boundary are such that $\mathbf{T} = \mathbf{T}^{(0)}$ which, using eqn. (9.8), implies that

$$\mathbf{T}^{(1)} = \mathbf{T}^{(2)} = 0. \tag{9.17}$$

Then it follows from eqn. (9.15) that $\mathbf{z} \equiv 0$ is a solution to eqn. (9.12). Therefore, there are only two unknown displacements, $\mathbf{v}$ and $\mathbf{w}$, to be solved for.

We note that $\mathbf{v}$ represents the solution in the special case where no scaling parameter is present and, as such, corresponds to the classical elasticity solution. Indeed, such a correspondence, in the case of planar problems, was shown to hold in Chapter 8 where several boundary value problems were solved. Once $\mathbf{v}$ is evaluated, it will serve to determine a body force term in the governing equation for $\mathbf{w}$ given in eqn. (9.13). It will also provide a surface traction term in the boundary condition for $\mathbf{w}$ as is evident from eqn. (9.16). Thus, the lower order solution will contribute to the determination of the higher order one. This approach has some resemblance to a method of solution in nonlinear elasticity, known as the method of successive approximations, due to Signorini (Green and Adkins 1970). However, what we have presented here is entirely within the context of doublet mechanics which makes it essentially different from the method of successive approximations.

Furthermore, not all the solutions at the second level of approximation are scale-dependent. For instance, when a doublet-mechanical body is subjected to a uniform hydrostatic pressure the displacement field $\mathbf{v}$ is linear in the spatial coordinates $\mathbf{x}$. As a result, the body force term in eqn. (9.13) and the surface traction term in eqn. (9.16) vanish, implying that $\mathbf{w} \equiv 0$ is a solution. This reasoning could be generalized to show that the terms involving higher powers of η, in a power series expansion of $\mathbf{u}$, also vanish, resulting in a solution which is independent of the scaling parameters.

9.3 Aspects of Well-Posedness of the Field Equations

One of the criteria that bears upon the question of well-posedness of a boundary value problem is existence of solutions. It is well known (Love 1944) that the necessary conditions for existence of equilibrium solutions, to a purely traction boundary value problem, are that the forces and moments acting on the body be self-equilibrated. That is,

$$\int_B \bar{\mathbf{F}}\, dx + \int_{\partial B} \bar{\mathbf{T}}\, ds = 0, \tag{9.18}$$

$$\int_B \mathbf{x} \times \bar{\mathbf{F}}\, dx + \int_{\partial B} \mathbf{x} \times \bar{\mathbf{T}}\, ds = 0, \tag{9.19}$$

where $\bar{\mathbf{T}}$ and $\bar{\mathbf{F}}$ are the prescribed surface tractions and body forces, respectively.

In the context of the present problem, with no prescribed body forces, we assume that $\mathbf{T}$ satisfies the above equations. Then, since eqn. (9.17) implies $\mathbf{T} = \mathbf{T}^{(0)}$, it is clear from eqns. (9.11) and (9.14) that the conditions for existence of $\mathbf{v}$ are automatically satisfied. Therefore, we only need to consider the corresponding conditions for $\mathbf{w}$. To this end, we rewrite eqns. (9.13) and (9.16), respectively, as

$$A_\circ \sum_{\alpha=1}^{n} \overset{\circ}{\tau}_{\alpha i}\, \overset{\circ}{\tau}_{\alpha j}\, \overset{\circ}{\tau}_{\alpha k}\, \overset{\circ}{\tau}_{\alpha l}\, w_{j,kl} + \bar{F}_i = 0, \tag{9.20}$$

$$A_\circ n_l \sum_{\alpha=1}^{n} \overset{\circ}{\tau}_{\alpha i}\, \overset{\circ}{\tau}_{\alpha j}\, \overset{\circ}{\tau}_{\alpha k}\, \overset{\circ}{\tau}_{\alpha l}\, w_{j,k} = \bar{T}_i, \tag{9.21}$$

$$\bar{F}_i = \frac{A_\circ}{12} \sum_{\alpha=1}^{n} \overset{\circ}{\tau}_{\alpha i}\, \overset{\circ}{\tau}_{\alpha j}\, \overset{\circ}{\tau}_{\alpha k}\, \overset{\circ}{\tau}_{\alpha l}\, \overset{\circ}{\tau}_{\alpha p}\, \overset{\circ}{\tau}_{\alpha q}\, v_{j,klpq}, \tag{9.22}$$

$$\bar{T}_i = \frac{1}{4} A_\circ n_l \sum_{\alpha=1}^{n} \overset{\circ}{\tau}_{\alpha i}\, \overset{\circ}{\tau}_{\alpha j}\, \overset{\circ}{\tau}_{\alpha k}\, \overset{\circ}{\tau}_{\alpha l}\, \overset{\circ}{\tau}_{\alpha p}\, \overset{\circ}{\tau}_{\alpha q}\, v_{j,kpq}. \tag{9.23}$$

Integrating eqn. (9.3) over the volume of the body and using the divergence theorem, we obtain

$$A_\circ \int_{\partial B} n_l \sum_{\alpha=1}^{n} \overset{\circ}{\tau}_{\alpha i}\, \overset{\circ}{\tau}_{\alpha j}\, \overset{\circ}{\tau}_{\alpha k}\, \overset{\circ}{\tau}_{\alpha l} \left(u_{j,k} + \frac{\eta^2}{12} \overset{\circ}{\tau}_{\alpha p}\, \overset{\circ}{\tau}_{\alpha q}\, u_{j,kpq} \right) ds = 0. \tag{9.24}$$

In addition, integrating eqn. (9.6) over the boundary of the body yields

$$A_\circ \int_{\partial B} n_l \sum_{\alpha=1}^{n} \overset{\circ}{\tau}_{\alpha i}\, \overset{\circ}{\tau}_{\alpha j}\, \overset{\circ}{\tau}_{\alpha k}\, \overset{\circ}{\tau}_{\alpha l} \left(u_{j,k} - \frac{\eta^2}{4} \overset{\circ}{\tau}_{\alpha p}\, \overset{\circ}{\tau}_{\alpha q}\, u_{j,kpq} \right) ds$$

$$= \int_{\partial B} T_i\, ds = 0, \tag{9.25}$$

where the last equality is the result of the assumption that **T** is a self-equilibrated vector field. Subtracting eqn. (9.25) from eqn. (9.24) results in an expression involving $u_{j,kpq}$ which, upon using eqn. (9.9), results in two equations corresponding to two different powers of η. The first of these equations is found to be

$$A_\circ \int_{\partial B} n_l \sum_{\alpha=1}^{n} \overset{\circ}{\tau}_{\alpha i}\, \overset{\circ}{\tau}_{\alpha j}\, \overset{\circ}{\tau}_{\alpha k}\, \overset{\circ}{\tau}_{\alpha l}\, \overset{\circ}{\tau}_{\alpha p}\, \overset{\circ}{\tau}_{\alpha q}\, v_{j,kpq}\, ds = 0. \tag{9.26}$$

From eqns. (9.23) and (9.26) it follows that

$$\int_{\partial B} \bar{T}_i\, ds = 0. \tag{9.27}$$

Then, upon using the divergence theorem again, the above expression implies that

$$\int_B \bar{F}_i\, dx = 0. \tag{9.28}$$

In view of eqns. (9.27) and (9.28), it is clear that eqn. (9.18) is satisfied and, as a result, the first necessary condition for existence of **w** is verified.

We verify the second existence condition, given in eqn. (9.19), in a similar fashion. First, we take the cross product of **x** with eqn. (9.6) and integrate the resulting expression over the boundary of the body to find

$$\begin{aligned} A_\circ \int_{\partial B} \epsilon_{sri}\, x_r\, n_l \sum_{\alpha=1}^{n} \overset{\circ}{\tau}_{\alpha i}\, \overset{\circ}{\tau}_{\alpha j}\, \overset{\circ}{\tau}_{\alpha k}\, \overset{\circ}{\tau}_{\alpha l} \left(u_{j,k} - \frac{\eta^2}{4}\, \overset{\circ}{\tau}_{\alpha p}\, \overset{\circ}{\tau}_{\alpha q}\, u_{j,kpq} \right) ds \\ = \int_{\partial B} \epsilon_{sri}\, x_r\, T_i\, ds = 0, \end{aligned} \tag{9.29}$$

where ϵ_{ijk} is the permutation tensor and the last equality follows from the assumption that **T** is a self-equilibrated vector field. Then, we take the cross product of **x** with eqn. (9.3) and integrate over the body to find

$$\begin{aligned} A_\circ \int_{B} \epsilon_{sri}\, x_r \sum_{\alpha=1}^{n} \overset{\circ}{\tau}_{\alpha i}\, \overset{\circ}{\tau}_{\alpha j}\, \overset{\circ}{\tau}_{\alpha k}\, \overset{\circ}{\tau}_{\alpha l} \left(u_{j,kl} + \frac{\eta^2}{12}\, \overset{\circ}{\tau}_{\alpha p}\, \overset{\circ}{\tau}_{\alpha q}\, u_{j,klpq} \right) dx \\ = A_\circ \int_{\partial B} \epsilon_{sri}\, x_r\, n_l \sum_{\alpha=1}^{n} \overset{\circ}{\tau}_{\alpha i}\, \overset{\circ}{\tau}_{\alpha j}\, \overset{\circ}{\tau}_{\alpha k}\, \overset{\circ}{\tau}_{\alpha l} \left(u_{j,k} + \frac{\eta^2}{12}\, \overset{\circ}{\tau}_{\alpha p}\, \overset{\circ}{\tau}_{\alpha q}\, u_{j,kpq} \right) ds \\ - A_\circ \int_{B} \epsilon_{sli} \sum_{\alpha=1}^{n} \overset{\circ}{\tau}_{\alpha i}\, \overset{\circ}{\tau}_{\alpha j}\, \overset{\circ}{\tau}_{\alpha k}\, \overset{\circ}{\tau}_{\alpha l} \left(u_{j,k} + \frac{\eta^2}{12}\, \overset{\circ}{\tau}_{\alpha p}\, \overset{\circ}{\tau}_{\alpha q}\, u_{j,kpq} \right) dx \quad (9.30) \\ = A_\circ \int_{\partial B} \epsilon_{sri}\, x_r\, n_l \sum_{\alpha=1}^{n} \overset{\circ}{\tau}_{\alpha i}\, \overset{\circ}{\tau}_{\alpha j}\, \overset{\circ}{\tau}_{\alpha k}\, \overset{\circ}{\tau}_{\alpha l} \left(u_{j,k} + \frac{\eta^2}{12}\, \overset{\circ}{\tau}_{\alpha p}\, \overset{\circ}{\tau}_{\alpha q}\, u_{j,kpq} \right) ds \\ = 0. \end{aligned} \tag{9.31}$$

Subtracting eqn. (9.31) from eqn. (9.29) we obtain

$$A_\circ \int_{\partial B} \epsilon_{sri}\, x_r\, n_l \sum_{\alpha=1}^{n} \overset{\circ}{\tau}_{\alpha i}\, \overset{\circ}{\tau}_{\alpha j}\, \overset{\circ}{\tau}_{\alpha k}\, \overset{\circ}{\tau}_{\alpha l}\, \overset{\circ}{\tau}_{\alpha p}\, \overset{\circ}{\tau}_{\alpha q}\, u_{j,kpq}\, ds = 0. \tag{9.32}$$

Substitution of eqn. (9.9) into eqn. (9.32) results in two equations corresponding to two different powers of η. The first of these equations is found to be

$$A_\circ \int_{\partial B} \epsilon_{sri}\, x_r\, n_l \sum_{\alpha=1}^{n} \overset{\circ}{\tau}_{\alpha i}\, \overset{\circ}{\tau}_{\alpha j}\, \overset{\circ}{\tau}_{\alpha k}\, \overset{\circ}{\tau}_{\alpha l}\, \overset{\circ}{\tau}_{\alpha p}\, \overset{\circ}{\tau}_{\alpha q}\, v_{j,kpq}\, ds = 0. \tag{9.33}$$

Comparison of eqn. (9.23) and eqn. (9.33) yields

$$\int_{\partial B} \epsilon_{sri}\, x_r\, \bar{T}_i\, ds = 0. \tag{9.34}$$

Finally, using the divergence theorem, the above implies that

$$\int_{B} \epsilon_{sri}\, x_r\, \bar{F}_i\, dx = 0. \tag{9.35}$$

It is evident from eqns. (9.34) and (9.35) that eqn. (9.19) is satisfied, thereby resulting in verification of conditions for existence of **w**. This establishes one of the main criteria for well-posedness of the boundary value problems which appear in the present formulation.

9.4 Kelvin's Problem in the Plane

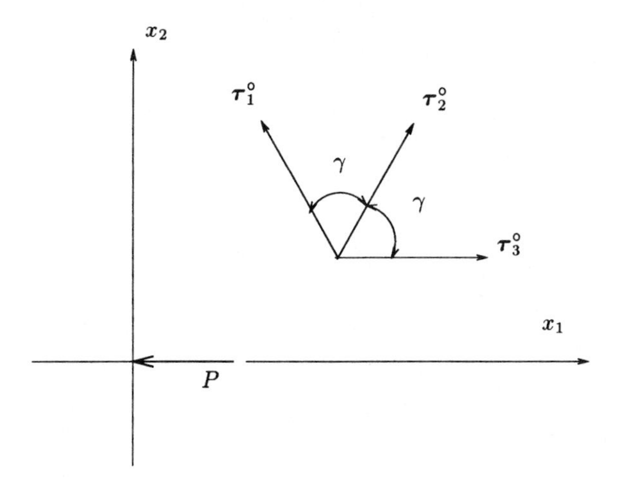

Fig. 9.1. Doublet geometry for Kelvin's problem

In order to illustrate an application of the present formulation in obtaining multi-scale solutions to boundary value problems, we solve Kelvin's problem in the plane. Namely, we consider a force of magnitude P acting at the origin of the plane in the direction of negative x_1-axis as shown in Fig. 9.1. We take the number of doublets to be three and their angle of separation to be $\pi/3$.

The governing equation for $\mathbf{v}$, obtained in eqn. (9.11), can be rewritten as

$$C_{ijkl}\, v_{k,lj} = 0, \tag{9.36}$$

where

$$C_{ijkl} = A_\circ \sum_{\alpha=1}^{n} \overset{\circ}{\tau}_{\alpha i}\, \overset{\circ}{\tau}_{\alpha j}\, \overset{\circ}{\tau}_{\alpha k}\, \overset{\circ}{\tau}_{\alpha l}, \tag{9.37}$$

subject to the boundary condition given in eqn. (9.14).

Equations (9.36) and (9.14) have the same form as the governing equations and boundary conditions for a traction boundary value problem in elasticity with the major difference that the elasticity tensor, given by eqn. (9.37), is a fourth order tensor possessing symmetry with respect to any permutation of its subscripts. On the other hand, for the particular geometry of the doublets considered here, $\mathbf{C}$ is an isotropic tensor. Then, the symmetries of the elasticity tensor, as shown in eqns. (5.69) and (5.70), result in the equality of Lame's constants for the doublet-mechanical body. That is, $\lambda = \mu$ which in turn gives the value of 1/4 for Poisson ratio. By equating the expression given in eqn. (9.37) to the one given in eqn. (5.69), it can be shown that $\lambda = 3A_\circ/8$.

Therefore, to find the solution to eqn. (9.36) we set the Lame's constants equal to each other in the elasticity solution of the planar Kelvin's problem as given in Love (1944, pp. 209–210). The result is

$$v_1 = c_1 \log r + c_2 \frac{x_2^2}{r^2}, \qquad v_2 = c_3 \frac{x_1 x_2}{r^2}, \tag{9.38}$$

where

$$c_1 = \frac{p}{3\pi\lambda}, \qquad c_2 = \frac{c_1}{2}, \qquad c_3 = -\frac{c_1}{2}. \tag{9.39}$$

The governing equation for $\mathbf{w}$, given in eqn. (9.20), is rewritten as

$$C_{ijkl}\, w_{k,lj} + \bar{F}_i = 0, \tag{9.40}$$

where

$$\bar{F}_i = D_{ij,j} = \frac{A_\circ}{12} \sum_{\alpha=1}^{n} \overset{\circ}{\tau}_{\alpha i}\, \overset{\circ}{\tau}_{\alpha j}\, \overset{\circ}{\tau}_{\alpha k}\, \overset{\circ}{\tau}_{\alpha l}\, \overset{\circ}{\tau}_{\alpha p}\, \overset{\circ}{\tau}_{\alpha q}\, v_{k,lpqj}, \tag{9.41}$$

and

$$D_{ij} = \frac{A_\circ}{12} \sum_{\alpha=1}^{n} \overset{\circ}{\tau}_{\alpha i}\, \overset{\circ}{\tau}_{\alpha j}\, \overset{\circ}{\tau}_{\alpha k}\, \overset{\circ}{\tau}_{\alpha l}\, \overset{\circ}{\tau}_{\alpha p}\, \overset{\circ}{\tau}_{\alpha q}\, v_{k,lpq}, \tag{9.42}$$

subject to the traction boundary conditions in eqn. (9.21).

We note that eqn. (9.40) has the same form as eqn. (9.36) with the additional presence of the body force $\bar{\mathbf{F}}$. To find $\mathbf{w}$, in analogy to what was done for $\mathbf{v}$, we solve the boundary value problem in elasticity subject to the body force $\bar{\mathbf{F}}$ and traction boundary conditions in eqn. (9.21) and equate the Lame's constants in the solution.

To this end, we first compute the tensor $\mathbf{D}$ as given in eqn. (9.42). On expanding eqn. (9.42), component D_{11} is found to be

$$
\begin{aligned}
D_{11} = \frac{A_\circ}{12} \big[& (\overset{\circ}{\tau}_{11})^2 \big((\overset{\circ}{\tau}_{11})^4 v_{1,111} + 2 (\overset{\circ}{\tau}_{11})^3 \overset{\circ}{\tau}_{12} v_{1,112} + (\overset{\circ}{\tau}_{11})^2 (\overset{\circ}{\tau}_{12})^2 v_{1,122} \\
& + (\overset{\circ}{\tau}_{11})^3 \overset{\circ}{\tau}_{12} v_{1,112} + 2 (\overset{\circ}{\tau}_{11})^2 (\overset{\circ}{\tau}_{12})^2 v_{1,122} + \overset{\circ}{\tau}_{11} (\overset{\circ}{\tau}_{12})^3 v_{1,222} \\
& + \overset{\circ}{\tau}_{12} (\overset{\circ}{\tau}_{11})^3 v_{2,111} + 2 (\overset{\circ}{\tau}_{11})^2 (\overset{\circ}{\tau}_{12})^2 v_{2,211} + \overset{\circ}{\tau}_{11} (\overset{\circ}{\tau}_{12})^3 v_{2,221} \\
& + (\overset{\circ}{\tau}_{11})^2 (\overset{\circ}{\tau}_{12})^2 v_{2,211} + 2 \overset{\circ}{\tau}_{11} (\overset{\circ}{\tau}_{12})^3 v_{2,221} + (\overset{\circ}{\tau}_{12})^4 v_{2,222} \big) \\
& + (\overset{\circ}{\tau}_{21})^2 \big((\overset{\circ}{\tau}_{21})^4 v_{1,111} + 2 (\overset{\circ}{\tau}_{21})^3 \overset{\circ}{\tau}_{22} v_{1,112} + (\overset{\circ}{\tau}_{21})^2 (\overset{\circ}{\tau}_{22})^2 v_{1,122} \\
& + (\overset{\circ}{\tau}_{21})^3 \overset{\circ}{\tau}_{22} v_{1,112} + 2 (\overset{\circ}{\tau}_{21})^2 (\overset{\circ}{\tau}_{22})^2 v_{1,122} + \overset{\circ}{\tau}_{21} (\overset{\circ}{\tau}_{22})^3 v_{1,222} \\
& + \overset{\circ}{\tau}_{22} (\overset{\circ}{\tau}_{21})^3 v_{2,111} + 2 (\overset{\circ}{\tau}_{21})^2 (\overset{\circ}{\tau}_{22})^2 v_{2,211} + \overset{\circ}{\tau}_{21} (\overset{\circ}{\tau}_{22})^3 v_{2,221} \\
& + (\overset{\circ}{\tau}_{21})^2 (\overset{\circ}{\tau}_{22})^2 v_{2,211} + 2 \overset{\circ}{\tau}_{21} (\overset{\circ}{\tau}_{22})^3 v_{2,221} + (\overset{\circ}{\tau}_{22})^4 v_{2,222} \big) \\
& + (\overset{\circ}{\tau}_{31})^2 \big((\overset{\circ}{\tau}_{31})^4 v_{1,111} + 2 (\overset{\circ}{\tau}_{31})^3 \overset{\circ}{\tau}_{32} v_{1,112} + (\overset{\circ}{\tau}_{31})^2 (\overset{\circ}{\tau}_{32})^2 v_{1,122} \\
& + (\overset{\circ}{\tau}_{31})^3 \overset{\circ}{\tau}_{32} v_{1,112} + 2 (\overset{\circ}{\tau}_{31})^2 (\overset{\circ}{\tau}_{32})^2 v_{1,122} + \overset{\circ}{\tau}_{31} (\overset{\circ}{\tau}_{32})^3 v_{1,222} \\
& + \overset{\circ}{\tau}_{32} (\overset{\circ}{\tau}_{31})^3 v_{2,111} + 2 (\overset{\circ}{\tau}_{31})^2 (\overset{\circ}{\tau}_{32})^2 v_{2,211} + \overset{\circ}{\tau}_{31} (\overset{\circ}{\tau}_{32})^3 v_{2,221} \\
& + (\overset{\circ}{\tau}_{31})^2 (\overset{\circ}{\tau}_{32})^2 v_{2,211} + 2 \overset{\circ}{\tau}_{31} (\overset{\circ}{\tau}_{32})^3 v_{2,221} + (\overset{\circ}{\tau}_{32})^4 v_{2,222} \big) \big] . \qquad (9.43)
\end{aligned}
$$

Given the doublet geometry in Fig. 9.1, the components of the doublet vectors are

$$\overset{\circ}{\tau}_{11} = -\frac{1}{2}, \quad \overset{\circ}{\tau}_{12} = \frac{\sqrt{3}}{2}, \quad \overset{\circ}{\tau}_{21} = \frac{1}{2}, \quad \overset{\circ}{\tau}_{22} = \frac{\sqrt{3}}{2}, \quad \overset{\circ}{\tau}_{31} = 1, \quad \overset{\circ}{\tau}_{32} = 0. \qquad (9.44)$$

Substituting eqn. (9.44) into eqn. (9.43) we find

$$D_{11} = \frac{A_\circ}{128} (11\, v_{1,111} + 3\, v_{1,122} + 3\, v_{2,211} + 3\, v_{2,222}). \qquad (9.45)$$

Other components of $\mathbf{D}$ are found in a similar manner. Omitting the details of the expansions, they are given as

$$D_{12} = \frac{A_\circ}{128} (3\, v_{1,112} + 3\, v_{1,222} + v_{2,111} + 9\, v_{2,221}), \qquad (9.46)$$

$$D_{22} = \frac{A_\circ}{128} (v_{1,111} + 9\, v_{1,122} + 9\, v_{2,211} + 9\, v_{2,222}). \qquad (9.47)$$

Now we can compute the components of the body force $\bar{\mathbf{F}}$. Using eqns. (9.38), (9.41), (9.45) and (9.46), the first component is given by

$$
\begin{aligned}
\bar{F}_1 &= \frac{A_\circ}{128} (11 v_{1,1111} + 6 v_{1,1122} + 4 v_{2,2111} + 3 v_{1,2222} + 12 v_{2,2221}) && (9.48) \\
&= \frac{3}{4} A_\circ c_2 \left(64 \frac{x_2^6}{r^{10}} - 104 \frac{x_2^4}{r^8} + 44 \frac{x_2^2}{r^6} - \frac{3}{r^4} \right). && (9.49)
\end{aligned}
$$

Similarly, the second component of the body force, using eqns. (9.38), (9.41), (9.46) and (9.47), is given by

$$\bar{F}_2 = \frac{A_\circ}{128}(4v_{1,1112} + 12v_{1,1222} + 18v_{2,2211} + v_{2,1111} + 9v_{2,2222}) \quad (9.50)$$

$$= 6\,A_\circ\,c_2 \left(2\,\frac{x_1 x_2}{r^6} - 9\,\frac{x_1 x_2^3}{r^8} + 8\,\frac{x_1 x_2^5}{r^{10}}\right). \quad (9.51)$$

The traction boundary conditions for $\mathbf{w}$, given in eqn. (9.21), can be rewritten as

$$A_\circ\, n_l \sum_{\alpha=1}^{n} \overset{\circ}{\tau}_{\alpha i}\, \overset{\circ}{\tau}_{\alpha j}\, \overset{\circ}{\tau}_{\alpha k}\, \overset{\circ}{\tau}_{\alpha l}\, w_{j,k} = 3\, D_{jl}\, n_l. \quad (9.52)$$

The components of $\mathbf{D}$, given in eqns. (9.45)–(9.47), vanish as r tends to infinity. Therefore, the tractions are zero at the boundary.

In order to obtain a solution for eqn. (9.40), we use Papkovich-Neuber displacement potentials denoted by ϕ and $\boldsymbol{\psi}$. When these potentials satisfy

$$\phi_{,jj} = -\frac{x_j\, F_j^*}{2(1-\nu)} = -\frac{2}{3}\, x_j\, \bar{F}_j, \quad (9.53)$$

$$\psi_{i,jj} = \frac{F_i^*}{2(1-\nu)} = \frac{2}{3}\, \bar{F}_i, \quad (9.54)$$

with Poisson ratio $\nu = 1/4$, the displacement field $\mathbf{w}$ is given by

$$w_i = \frac{4}{3A_\circ}\,(\phi_{,i} + x_j\, \psi_{j,i} - 2\,\psi_i). \quad (9.55)$$

Given the body force field in eqns. (9.49)–(9.51), particular solutions for eqns. (9.53) and (9.54) are found to be

$$\phi = -2\, A_\circ\, c_2 \left(\frac{1}{48}\,\frac{x_1}{r^2} - \frac{3}{8}\,\frac{x_1 x_2^2}{r^4} + \frac{x_1 x_2^4}{r^6} - \frac{2}{3}\,\frac{x_1 x_2^6}{r^8}\right), \quad (9.56)$$

$$\psi_1 = \frac{1}{2}\, A_\circ\, c_2 \left(\frac{1}{8r^2} - \frac{7}{4}\,\frac{x_2^2}{r^4} + \frac{11}{3}\,\frac{x_2^4}{r^6} - 2\,\frac{x_2^6}{r^8}\right), \quad (9.57)$$

$$\psi_2 = A_\circ\, c_2 \left(-\frac{3}{8}\,\frac{x_1 x_2}{r^4} + \frac{4}{3}\,\frac{x_1 x_2^3}{r^6} - \frac{x_1 x_2^5}{r^8}\right). \quad (9.58)$$

Substituting eqns. (9.56)–(9.58) into eqn. (9.55), we find the displacement field $\mathbf{w}$ to be

$$w_1 = -\frac{c_2}{18}\left(128\,\frac{x_2^8}{r^{10}} - 352\,\frac{x_2^6}{r^8} + 320\,\frac{x_2^4}{r^6} - 100\,\frac{x_2^2}{r^4} + \frac{5}{r^2}\right), \quad (9.59)$$

$$w_2 = \frac{c_2}{9}\, x_1 \left(64\,\frac{x_2^7}{r^{10}} - 96\,\frac{x_2^5}{r^8} + 32\,\frac{x_2^3}{r^6} + \frac{x_2}{r^4}\right). \quad (9.60)$$

Thus, the displacement field of the original problem is expressed as

$$\mathbf{u} = \mathbf{v} + \eta^2 \, \mathbf{w}, \tag{9.61}$$

where $\mathbf{v}$ is given by eqn. (9.38) and $\mathbf{w}$ is given by eqns. (9.59) and (9.60). The displacement field $\mathbf{u}$ represents the doublet-mechanical solution of the planar Kelvin's problem, at the second level of approximation, to within the accuracy of terms of order η^3.

9.5 Discussion

Having determined the displacements $\mathbf{v}$ and $\mathbf{w}$, other field quantities such as the microstresses and the macrostresses can be readily evaluated. We recall that the elongation microstress $p_{\alpha i}$ is given by eqn. (9.4). The relationship between the microstresses and the macrostresses was developed in eqn. (1.40). Writing this expression, at the second level of approximation, yields

$$\sigma_{ji} = \sigma_{ji}^{(2)} = \sum_{\alpha=1}^{n} \overset{\circ}{\tau}_{\alpha j} \left(p_{\alpha i} - \frac{\eta}{2} \, \overset{\circ}{\tau}_{\alpha k} \, \frac{\partial p_{\alpha i}}{\partial x_k} \right) . \tag{9.62}$$

Now, both the microstresses and the macrostresses can be written in terms of the displacements. When eqn. (9.9) is substituted into eqn. (9.4), and terms of order η^3 are neglected, we find

$$p_{\alpha i} = A_\circ \, \overset{\circ}{\tau}_{\alpha i} \, \overset{\circ}{\tau}_{\alpha j} \, \overset{\circ}{\tau}_{\alpha k} \left(v_{j,k} + \frac{\eta}{2} \, \overset{\circ}{\tau}_{\alpha l} \, v_{j,kl} + \eta^2 \, w_{j,k} \right) . \tag{9.63}$$

Similarly, when eqn. (9.63) is substituted into eqn. (9.62) we find, to within terms of order η^3, that

$$\sigma_{ji} = A_\circ \sum_{\alpha=1}^{n} \overset{\circ}{\tau}_{\alpha j} \, \overset{\circ}{\tau}_{\alpha i} \, \overset{\circ}{\tau}_{\alpha k} \, \overset{\circ}{\tau}_{\alpha l} \left(v_{k,l} + \eta^2 \, w_{k,l} - \frac{\eta^2}{4} \, \overset{\circ}{\tau}_{\alpha p} \, \overset{\circ}{\tau}_{\alpha q} \, v_{k,lpq} \right) . \tag{9.64}$$

As expected, the displacement field $\mathbf{w}$ has a singularity at the origin due to the application of the point force. This singularity, however, is stronger than the one shown by $\mathbf{v}$ since the former is of $1/r^2$ type, whereas the latter is of the type $\log r$. It is for this reason that $\mathbf{v}$ grows unboundedly as $r \to \infty$, while $\mathbf{w}$ approaches zero asymptotically.

Although, in the foregoing development we determined an expansion of $\mathbf{u}$ up to the terms involving second power of η, the formulation can be extended in an obvious way to any arbitrary power of η. The characteristic feature of this formulation is that the lower order terms in the expansion of $\mathbf{u}$ contribute to the determination of the higher order terms. Thus, the displacement field in doublet mechanics, at the second level of approximation, can be determined to within any desirable accuracy in the terms involving the scaling parameter.

10. A New Direction: Nanotubes

Nasreen G. Chopra and A. Zettl

10.1 Introduction

The recent discovery of various stoichiometries of $B_xC_yN_z$ nanotubes provides an interesting application for the formalism covered in the previous chapters. A nanotube can be described as a long thin strip, cut out of a single atomic plane of material, rolled to form a cylinder with a diameter of nanometer scale and a length on the order of microns. This chapter will discuss the discovery of tubes of various layered materials, give a theoretical formulation of the cylindrical nanostructures, show some experimental work done on the mechanical behavior of carbon tubes and provide possible applications of this material in the future.

10.2 Discovery

Fig. 10.1. Transmission electron micrograph of a single-wall carbon nanotube. Line drawing shows the cross section of the structure.

Carbon nanotubes ($x = 0$, $y = 1$, $z = 0$) were first discovered by Iijima (1991) while performing transmission electron microscopy (TEM) on a fullerene sample taken from the chamber where C_{60} is produced. Fig. 10.1 is a micrograph

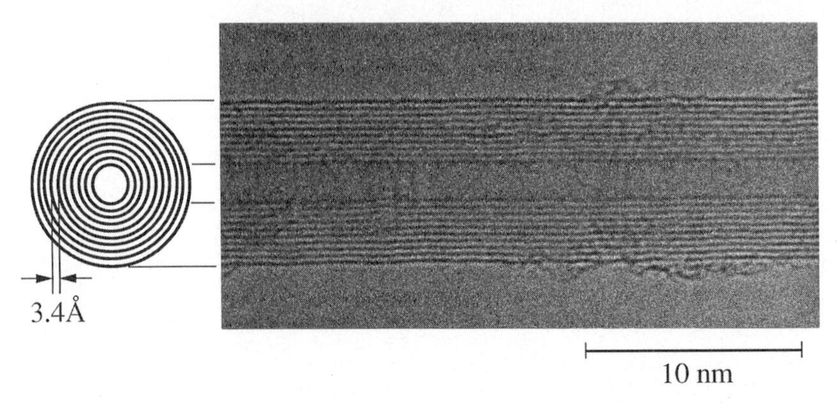

Fig. 10.2. Transmission electron micrograph of a multi-wall carbon nanotube. Line drawing shows the cross section of the tube, emphasizing the relative positions between the walls.

of a single-wall carbon nanotube. The dark line contrast comes from the single atomic layer thickness of the tube wall while the gap, which is 1.28nm, represents the inner diameter. Carbon nanotubes can also have more than one wall. Multi-wall nanotubes are hollow cylindrical structures with several concentric layers, as seen in Fig. 10.2. On either side of the gap, there is an equal number of lattice fringes indicating the number of walls of the tube. This particular multi-walled tube, for example, has 10 walls and an inner tube diameter of 1.6nm. Tubes with over 30 walls have been observed. However, the spacing between the multiple tube layers is always about the same as the interplanar distance in graphite, ~ 3.4Å, as pointed out in Fig. 10.2.

Inner diameters of single-wall and multi-walled carbon nanotubes range from 0.7 to 1.5nm and 1.3 to 4nm, respectively, while their lengths are commonly on the order of microns. Fig. 10.3 is a lower magnification TEM micrograph of a multi-walled carbon nanotube sample with several tubes which have lengths exceeding a micron thus leading to aspect ratios of over a thousand. The tubes are closed structures, capped off with some arrangement of carbon atoms which involves pentagon or heptagon formation along with the rest of the hexagonal network. It is possible, however to open and fill the tubes as Ajayan and Iijima (1993) have shown.

Carbon nanotubes are synthesized by arcing a graphite rod against a cooled electrode in a helium atmosphere. Although the growth kinetics are not well understood for the uncontrolled dynamics of the arc-discharge method, Ebbesen and Ajayan (1992) found the optimum parameters for multi-walled tube growth yielding approximately 50% tubes and the rest of the graphite turning into other layered nanostructures and nanoparticles of carbon. Single-wall tube growth results when a metal catalyst such as iron or cobalt is introduced in the arc-discharge chamber, as shown by Iijima and Ichihashi (1993) and Bethune et al (1993). Since these initial studies, various other techniques

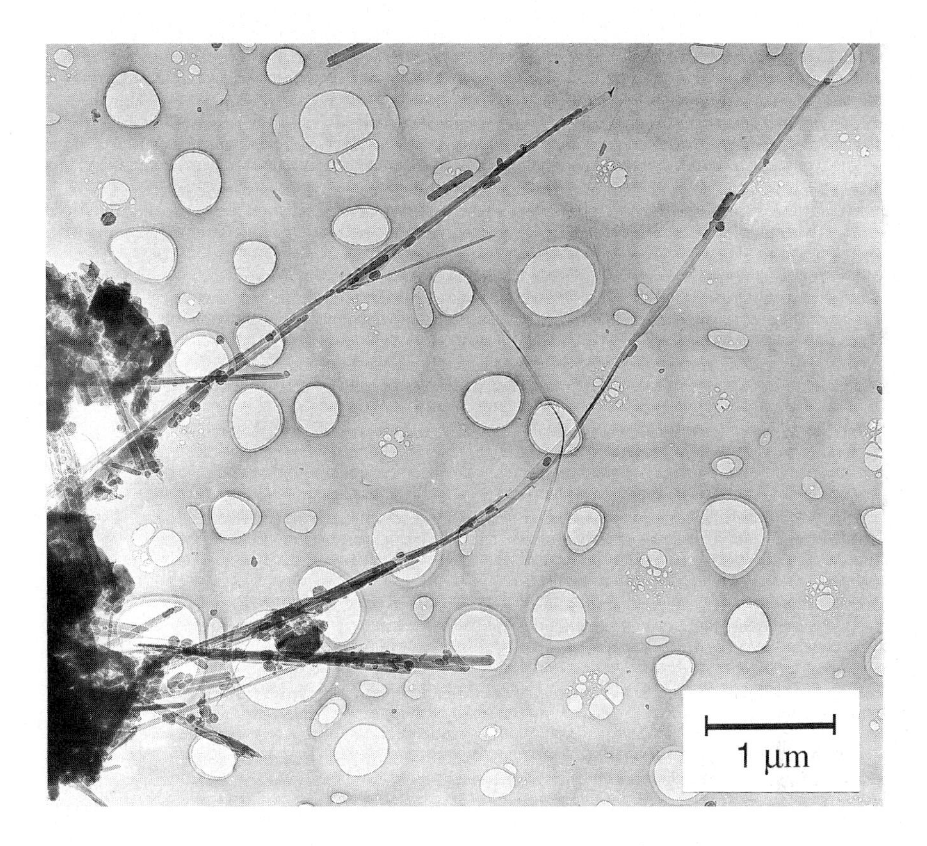

Fig. 10.3. TEM image of multi-wall carbon nanotubes on a holey carbon grid which is used to support the structures for TEM analysis. In general, nanotubes have high aspect ratios. This image also shows the layered spherical nanostructures that are produced while synthesizing carbon nanotubes.

for both single and multi-walled tube growth have been successful, including vapor condensation (Ge and Sattler 1993), laser ablation (Guo et al 1995) and electrolytic method (Hsu et al 1995).

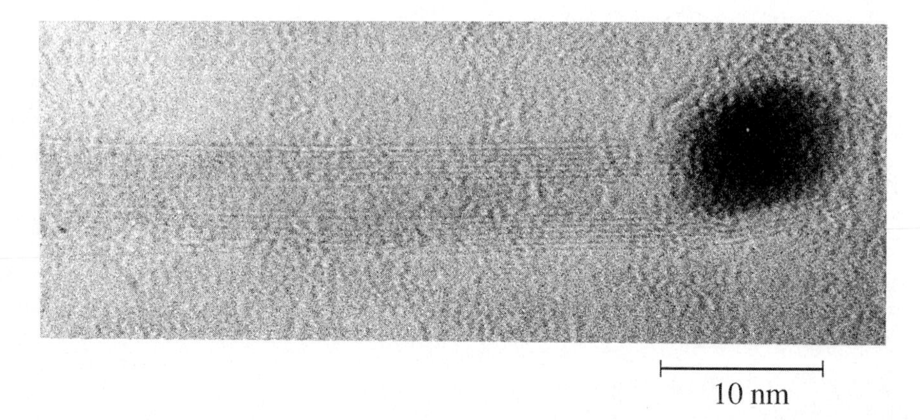

Fig. 10.4. TEM image of multi-wall BN nanotube produces by arcing a BN-filled tungsten electrode. The metal particle seen at the end is believed to help nucleate or terminate tube growth.

Weng-Sieh et al (1995) and Stephan et al (1994) synthesized tubes with $x = 1$, $y = 2$, $z = 1$ stoichiometry of boron, carbon, and nitrogen, namely BC_2N. Most recently boron nitride ($x = 1$, $y = 0$, $z = 1$) nanotubes have been discovered both multi-walled (Chopra et al 1995b) and single-walled (Loiseau et al 1996). All varieties of $B_xC_yN_z$ nanotubes which exist so far have been formed in the arc-discharge chamber using different materials and operating conditions. Fig. 10.4 (Chopra et al 1995b) shows a multi-walled BN nanotube produced by using a BN filled tungsten electrode; tubes synthesized by this method can have a metal particle at the tip which is believed to aid in nucleation or termination of BN tube growth.

10.3 Theory

The theoretical approach to nanotubes begins with defining indices relative to the lattice vectors on the hexagonal plane from which the tubes are formed. Fig. 10.5 shows the simplest hexagonal lattice made from a homogeneous material, simulating the case for carbon. The other, more complicated, materials will have bigger unit cells but can be treated with a similar approach because they are also planar, hexagonal materials in bulk. However, for the sake of simplicity, we consider the case where each vertex is occupied by the same kind of atom, for example carbon atoms in graphite. A tube is formed

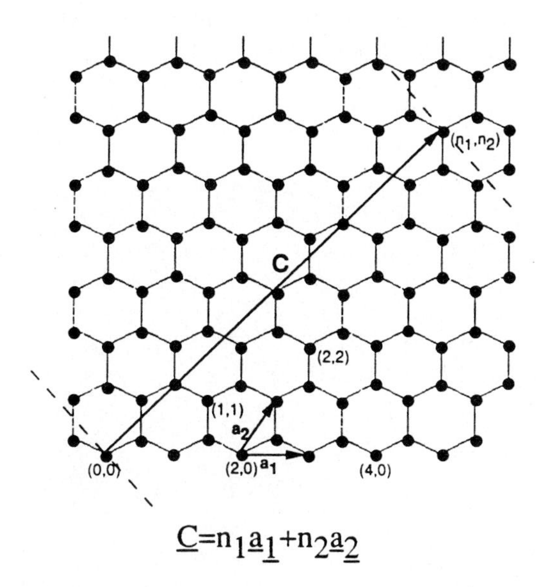

Fig. 10.5. Hexagonal network of atoms defined by indices relative to the lattice vectors. By cutting a strip (dashed line in figure) of this hexagonal sheet and mapping the origin to some (n_1, n_2), a tube is formed, uniquely defined by its circumference vector, **C**.

by defining a particular circumference vector $\mathbf{C} = n_1\,\mathbf{a}_1 + n_2\,\mathbf{a}_2$ where n_1 and n_2 are integers and $\mathbf{a}_1$ and $\mathbf{a}_2$ are the unit vectors (the same notation as used by Saito et al (1992)); having defined the width of the strip with **C**, we can form the tube by mapping the origin to the point (n_1, n_2) and the corresponding points down the length of the tube. Thus tubes will have several different chiralities, depending on the circumference vector, **C**, as seen in the examples of Fig. 10.6a-c.

Calculations of the density of states for carbon nanotubes indicate that the electrical properties of the tubes will range from semi-conducting to metallic depending on the chirality and the diameter of the tube (Saito et al 1992). Miyamoto et al (1994) have performed similar calculation of the electrical properties for BC_2N tubes and have found that they also range from semi-conducting to metallic like carbon nanotubes, but all BN nanotubes are predicted to be semi-conducting independent of their chirality and diameter (Blase et al 1994).

10.4 Mechanical Behavior

The unique size of these nanostructures leads to interesting questions about their mechanical behavior. Can the tubes be treated as continuos hollow

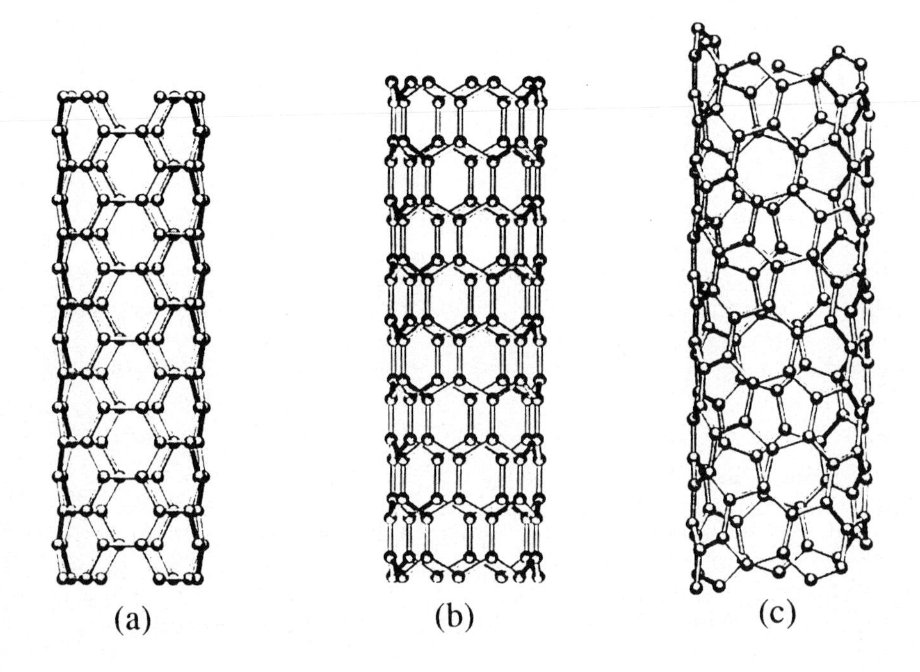

Fig. 10.6. Theoretical examples of tubes with different chiralities. (a) (4,4) armchair chirality tube (b) (8,0) zigzag tube (c) (7,2) tube. The names for the chiralities come from the arrangement of atoms at the cross-sectional edge of the tube as seen in the figure. (Courtesy of Vincent H. Crespi)

cylindrical structures or does their nanometer size call for a more discrete treatment? A doublet mechanical analysis of the mechanical properties of nanotubes has the potential to answer such a question. Meanwhile, energetics and Hamiltonian formulation predicts that carbon nanotubes will be 30 times stiffer than a steel rod of the same dimension (Ouverney et al 1992).

Much of the experimental work carried out on nanotubes has concentrated on carbon nanotubes. Analysis of high resolution TEM micrographs of some of these structures reveals defects in the cylindrical shape of the tube. Hiura et al (1994) found tube cross-sections that were slightly elliptical, and Ruoff et al (1993) have seen tubes with deformed walls which locally change the inner diameter by a small amount. Tubes with localized kinks and bends have also been observed in carbon nanotubes (Endo et al 1993, Despres et al 1995). We expect, then, in following with analogies of hollow cylindrical structures, to see nanotubes which have totally collapsed. Indeed, such structures have been observed and characterized for carbon nanotubes and an example is given in Fig. 10.7 (Chopra et al 1995a). On initial inspection collapsed carbon nanotubes resemble a ribbon-like structure with two distinct regions: a flat region where the wide part of the flattened tube is in the image plane and a twist region where the wide region is perpendicular to the imaging plane. High resolution TEM images of these two regions reveal the following: when the flat part is in the imaging plane two parallel sets of equal-numbered lattice fringes, separated by a gap are observed; and when the width of the ribbon is perpendicular to the image plane the same total number of lattice fringes are seen but with no gap. These features are characterized in the insets in Fig. 10.7 in order to give a clear representation of the geometries. Careful analysis of the structure including an in situ rotation study proved that the projected diameter of the collapsed structure changed as the sample was rotated in contrast to the conventional inflated tube where, as expected, the projection did not change with rotation of the sample.

10.5 Applications

The study of nanotubes is of interest to a variety of researchers from different fields mainly because of the potentially wide reaching effect these structures could have on various areas of technology. The most exciting and realizable potential from the electronics industry is to look at single-wall nanotubes, some of which are predicted to being metallic, as nanowires. The forefront of the microchip industry produces micron scale circuitry; nanotubes boast the ability to reduce the scale by a factor of a thousand. In fact, due to the semiconducting nature of BN nanotubes, theoretically, it is possible to dope with different substances to achieve p-type and n-type doping leading to the eventual revolutionary idea of having whole circuits on single nanotubes.

Also the predicted high strength, high flexibility of these nanostructures makes them candidates for building blocks of extremely strong, versatile ma-

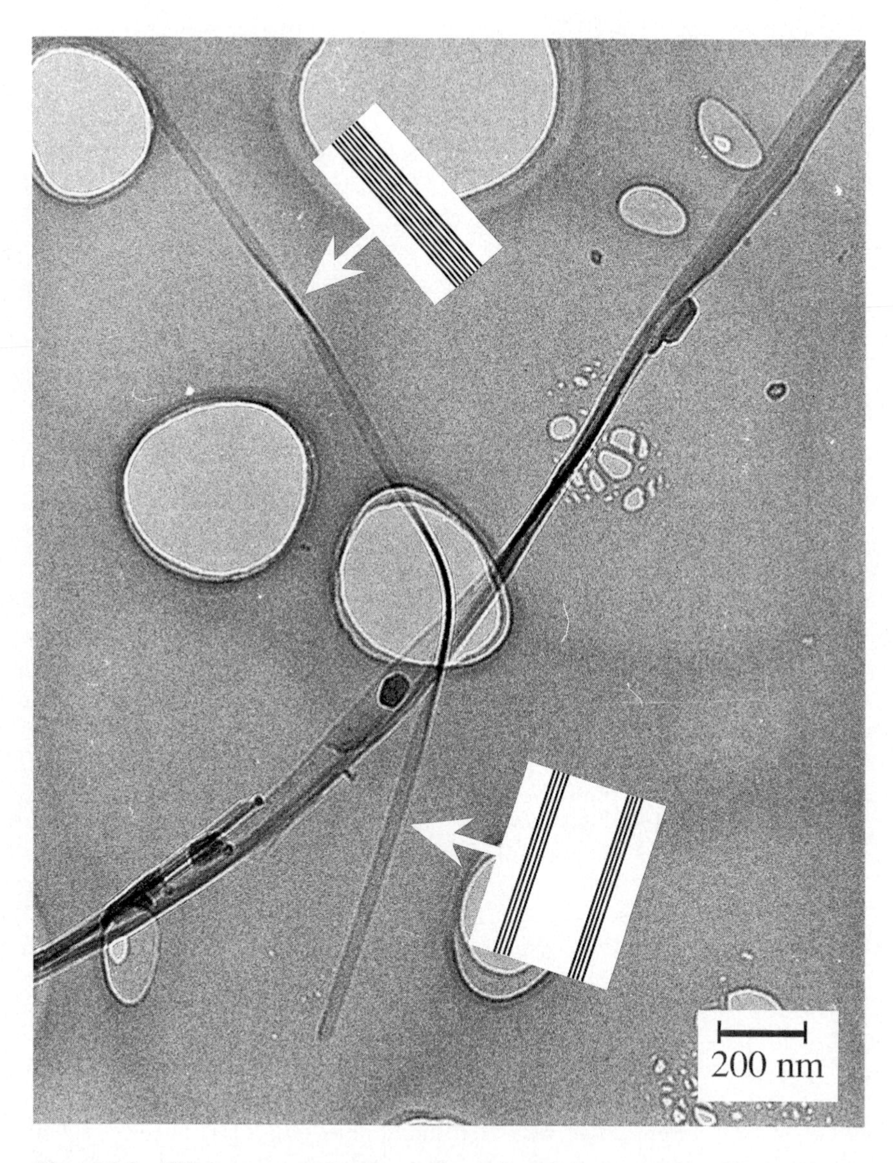

Fig. 10.7. TEM micrograph of a collapsed carbon nanotube. The insets (not to scale) emphasize the unique features of a flattened multi-wall nanotube. This ribbon-like structure has two distinct regions: a flat part, as illustrated in the lower inset, has an equal number of lattice fringes on either side of the gap, while the twist part has the same total number of lattice fringes but no gap (top inset).

terials. Presently graphite fibers are widely used in industry for their high performance; carbon nanotubes with their mostly defect free, crystalline walls are likely to be a vast improvement upon them.

References

Achenbach, J. D. (1980). *Wave Propagation in Elastic Solids*, North-Holland, Amsterdam.

Achenbach, J. D. (1993). *Evaluation of Materials and Structures by Quantitative Ultrasonic*, Springer-Verlag, Vienna.

Aero, E. L. and Kuvshinski, E. V. (1960). "Fundamental equations of the theory of elastic media with rotationally interacting particles." *Soviet Physics Solid State*, 2:1272–1281.

Ajayan, P. M. and Iijima, S. (1993). "Capillarity-induced filling of carbon nanotubes." *Nature*, 361:333.

Anderson, D. L., Minster, B., and Cole, D. (1974). "The effect of oriented cracks on seismic velocities." *J. Geophys. Res.*, 79:4011–4015.

Ashcroft, N. W. and Mermin, N. D. (1976). *Solid State Physics*, Holt, Rhinehart and Winston Inc., New York.

Askar, A. (1985). *Lattice Dynamical Foundations of Continuum Theories*, World Scientific Publishing Ltd., Singapore.

Båth, M. (1968). *Mathematical Aspects of Seismology*, Elsevier, Amsterdam.

Bacon, G. (1962). *Neutron Diffraction*, Clarendon Press, Oxford.

Barker, Jr., A. S., Merz, J. L., and Gossard, A. C. (1978). "Study of zone-folding effects on phonons in monolayers of GaAs-AlAs." *Phys. Review*, 17:3181–3196.

Bedford, A. and Drumheller, D. S. (1994). *Introduction to Elastic Wave Propagation*, John Wiley and Sons, Chichester, England.

Ben-Menahem, A. and Singh, S. J. (1981). *Seismic Waves and Sources*, Springer-Verlag, New York.

Beranek, W. J. and Hobbelman, G. J. (1992). "A mechanical model for brittle materials." In *6th Canadian Masonry Symposium*, pages 1–12, Saskatchewan, Canada, University of Saskatchewan.

Bethune, D. S., Kiang, C. H., de Vries, M. S., Gorman, G., Savoy, R., Vazquez, J., and Beyers, R. (1993). "Cobalt-catalysed growth of carbon nanotubes with single-atomic-layer walls." *Nature*, 363:605–607.

Blakemore, M. and Georgiou, G. A. (1988). *Mathematical Modelling in Non-Destructive Testing*, Clarendon Press, Oxford.

Blase, X., Rubio, A., Louie, S. G., and Cohen, M. L. (1994). "Stability and band gap constancy of boron nitride nanotubes." *Europhysics Letters*, 28:335–340.

Born, M. and Huang, K. (1962). *Dynamical Theory of Crystal Lattices*, Clarendon Press, Oxford.

Born, M. and Oppenheimer, J. R. (1927). "Zur quantumtheorie der molekeln." *Ann. Phys.*, 84:417–484.

Born, M. and von Kármán, T. (1912). "Über schwingungen in raumgittern." *Phys. Zeitschr.*, 13:297–309.

Böttger, H. (1983). *Principles of the Theory of Lattice Dynamics*, Physik-Verlag, Weinheim.

Brand, L. (1966). *Differential and Difference Equations*, John Wiley and Sons, New York.

Brekhovskikh, L. M. (1980). *Waves in Layered Media*, Academic Press, New York.

Brillouin, L. (1960). *Wave Propagation and Group Velocity*, Academic Press, New York.

Brillouin, L. (1963). *Wave Propagation in Periodic Structures*, Dover, New York.

Bromwich, T. (1898). "On the influence of gravity on elastic waves and, in particular, on the vibrations of an elastic globe." *Proc. London Math. Soc.*, 30:98–120.

Buchwald, V. T. and Davis, A. (1963). "Surface waves in elastic media with cubic symmetry." *Quart. J. Mech. and Appl. Mech.*, 16:283–294.

Casey, J. and Naghdi, P. (1981). "An invariant infinitesimal theory of motions superposed on a given motion." *Arch. Rat. Mech. Anal.*, 76:355–391.

Catlow, C. R. A., Dixon, M., and Mackrodt, W. C. (1982). *Computer Simulation of Solids*, chapter Interionic Potential in Ionic Solids, pages 130–161. Springer-Verlag, Berlin.

Chadwick, P. and Smith, G. D. (1982). "Surface waves in cubic elastic materials." In Hopkins, H. G. and Sewell, M. J., editors, *Mechanics of Solids. The Rodney Hill 60th Anniversary Volume*, pages 47–100, Pergamon, Oxford.

Chopra, N. G., Benedict, L. X., Crespi, V. H., Cohen, M. L., Louie, S. G., and Zettl, A. (1995). "Fully collapsed carbon nanotubes." *Nature*, 14:135–138.

Chopra, N. G., Luyken, R, J., Cherrey, K., Crespi, V. H., Cohen, M. L., Louie, S. G., and Zettl, A. (1995). "Boron nitride nanotubes." *Science*, 269:966–967.

Cochran, W. (1963). "Interpretation of phonon dispersion curves." In Wallis, R. F., editor, *Proc. Int. Conf. Lattice Dynamics*, pages 75–84, Oxford, Pergamon Press.

Cochran, W. (1973). *The Dynamics of Atoms in Crystals*, William Clowes and Sons Ltd., London.

Coleman, B. D. and Noll, W. (1959). "On the thermostatics of continuous media." *Arch. Rat. Mech. Anal.*, 4:99–128.

Coleman, B. D. and Noll, W. (1963). "The thermodynamics of elastic materials with heat conduction and viscosity." *Arch. Rat. Mech. Anal.*, 13:167–178.

Cosserat, E. and Cosserat, F. (1907). "Sur la mécanique générale." *Comptes Rendus, Acad. Sci., Paris*, 145:1139–1142.

Cosserat, E. and Cosserat, F. (1909). *Theorie des Corps Deformables*, Hermann, Paris.

Cottrell, A. H. (1964). *The Mechanical Properties of Matter*, John Wiley, New York.

Crampin, S. and Taylor, D. B. (1971). "The propagation of surface waves in anisotropic dedia." *Geophys. J.*, 25:71–87.

Crampin, S. (1981). "A review of wave motion in the anisotropic and cracked elastic media." *Wave Motion*, 3:343–391.

Crampin, S. (1984). "Effective anisotropic elastic constants for wave propagating through cracked solids." *Geophys. J. Royal Astronom. Soc.*, 76:135–145.

Cundall, P. A. and Strack, O. D. L. (1979). "A discrete numerical model for granular assemblies." *Geotechnique*, 29(1):47–65.

Dahler, J. S. and Scriven, L. E. (1963). "Theory of structured continua." *Proc. Phil. Soc. London, Ser. A*, 275:504–527.

Datta, S. K., Achenbach, J. D., and Rajapakse, Y. S. (1990). *Elastic Waves and Ultrasonic Nondestructive Evaluation*, North-Holland, Amsterdam.

Day, W. (1977). "An objection to using entropy as a primitive concept in continuum thermodynamics." *Acta Mech.*, 27:251–255.

DeGroot, S. R. and Mazur, P. (1962). *Non-Equilibrium Thermodynamics*, North Holland Publishing Company, Amsterdam.

Deresiewicz, H. (1958). *Mechanics of Granular Matter*, pages 233–306. Academic Press, New York.

Despres, J. F., Daguerre, E., and Kafdi, K. (1995). "Flexibility of graphene layers in carbon nanotubes." *Carbon*, 33(1):87–92.

Djafari-Rouhani, B., Maradudin, A. A., and Wallis, R. F. (1984). "Rayleigh waves on a superlattice stratified normal to the surface." *Phys. Review*, 29:6454–6462.

Dobrzynski, L., Djafari-Rouhani, B., and Hardouin Dupare, O. (1984). "Theory of surface phonons in superlattices." *Phys. Review*, 29:3138–3147.

Ebbesen, T. W. and Ajayan, P. M. (1992). "Large-scale synthesis of carbon nanotubes." *Nature*, 358:220–222.

Edelen, D. G. B., Green, A. E., and Laws, N. (1971). "Nonlocal continuum mechanics." *Arc. Rat. Mech. Anal.*, 43:36–44.

Edelen, D. G. B. (1976). *Continuum Physics*, volume 4, chapter Nonlocal Field Theories, pages 76–204. Academic Press, New York.

Elliott, R. J. and Gibson, A. F. (1974). *An Introduction to Solid State Physics and its Applications*, Harpers and Row Publishers, Inc., New York.

Endo, M., Takeuchi, K., Igarashi, S., Kobori, K., Shiraishi, M., and Kroto, H. W. (1993). "The production and structure of pyrolytic carbon nanotubes." *Journal of Physics and Chemistry of Solids*, 54:1841–1848.

Ericksen, J. L. and Truesdell, C. (1958). "Exact theory of stress and strain in rods and shells." *Arc. Rat. Mech. Anal.*, 1:295–323.

Eringen, A. C. and Edelen, D. G. B. (1972). "On non-local elasticity." *International Journal of Engineering Science*, 10:233–248.

Eringen, A. C. and Suhubi, E. S. (1964). "Nonlinear theory of simple microelastic solids I." *International Journal of Engineering Science*, 2:189–203.

Eringen, A. C. (1966). "Linear theory of micropolar elasticity." *J. Math. Mech.*, 15:909–924.

Eringen, A. C. (1968). "Mechanics of micromorphic continua." In Kröner, E., editor, *Proc. IUTAM Symposium Mechanics of Generalized Continua*, pages 18–35, Berlin, Sppringer-Verlag.

Eringen, A. C. (1973). "Linear theory of nonlocal microelasticity and dispersion of plane waves." *Lett. Appl. Sci.*, 1:129–146.

Ewing, W. M., Jardetzky, W. S., and Press, F. (1957). *Elastic Waves in Layered Media*, McGraw-Hill, New York.

Ferrari, M. and Granik, V. T. (1994). "Doublet-based micromechanical approaches to yield and failure criteria." *Material Science and Engng.*, A175:21–29.

Ferrari, M. and Granik, V. T. (1995). "Ultimate criteria for materials with different properties in tension and compression: A doublet-mechanical approach." *Material Science and Engng.*, A202(1–2):84–93.

Ferrari, M. (1985). "Termomeccanica dei mezzi continui classici e micromorfici." Tesi in Matematica, University of Padova, Italy.

Fraser, L. N. (1990). "Dynamic elasticity of microbedded and fractured rocks." *J. Geophys. Res.*, 95:4821–4831.

Fung, Y. C. (1977). *A First Course in Continuum Mechanics*, Prentice-Hall, Englewood Cliffs, New Jersey.

Garbin, H. D. and Knopoff, L. (1975). "Elastic moduli of a medium with liquid-filled cracks." *Quart. Appl. Math.*, 33:301–303.

Gassmann, F. (1951). "Elastic waves through a packing of spheres." *Geophysics*, 16:673–685.

Gauthier, R. D. and Jahsman, W. E. (1975). "A quest for micropolar elastic constants." *Journal of Applied Mechanics*, 42:369–374.

Gazis, D. C., Herman, R., and Wallis, R. F. (1960). "Surface elastic waves in cubic crystals." *Phys. Review*, 119:533–544.

Ge, M. and Sattler, K. (1993). "Vapor-condensation generation and STM analysis of fullerene tubes." *Science*, 260(23):515–518.

Gladwell, G. M. L. (1980). *Contact Problems in the Classical Theory of Elasticity*, Sijthoff and Noordhoff, The Netherlands.

Goodier, J. N. and Bishop (1951). "A note on critical reflection of elastic waves at free surfaces." *Journal of Applied Physics*, 23:124–126.

Goodman, M. A. and Cowin, S. C. (1972). "A continuum theory for granular materials." *Archive for Rational Mechanics and Analysis*, 44:249.

Gould, P. L. (1983). *Introduction to a Linear Elasticity*, Springer-Verlag, Berlin.

Granik, V. and Ferrari, M. (1993). "Microstructural mechanics of granular media." *Mechanics of Materials*, 15:301–322.

Granik, V. T. and Ferrari, M. (1994). "Micromechanical theory of yield and failure criteria for isotropic materials in view of hydrostatic stress components." Technical Report SEMM Report 94–21, Department of Civil Engineering, University of California at Berkeley, Berkeley.

Granik, V. and Ferrari, M. (1996). "Effects of long-range nonlocality on the tension of elastic strings." Technical Report SEMM 96-6, Department of Civil and Environmental Engineering, University of California at Berkeley, Berkeley,CA.

Granik, V. (1978). "Microstructural mechanics of granular media." Technical Report IM/MGU 78-241, Institute of Mechanics of Moscow State University. In Russian.

Green, A. E. and Adkins, J. E. (1970). *Large Elastic Deformations*, Oxford University Press, second edition.

Green, A. E. and Naghdi, P. M. (1965). "A dynamical theory of interacting continua." *International Journal of Engineering Science*, 3:231.

Green, A. E. and Naghdi, P. M. (1977). "On thermodynamics and the nature of the second law." *Proc. R. Soc. Lond. A*, 357:253–270.

Green, A. E. and Naghdi, P. M. (1978). "On thermodynamics and the nature of the second law for mixtures of interacting continua." *Quarterly Journal of Mathematics and Applied Mechanics*, 31(3):265–293.

Green, A. E. and Naghdi, P. M. (1979). "A note on invariance under superposed rigid body motions." *J. Elas.*, 9(1):1–8.

Green, A. E. and Rivlin, R. S. (1964). "Multipolar continuum mechanics." *Arch. Ratl. Mech. Analysis*, 17:113–147.

Green, A. E. and Rivlin, R. S. (1964). "Simple force and stress multipoles." *Arch. Ratl. Mech. Analysis*, 16:325–353.

Green, A. E. (1965). "Micro-materials and multipolar continuum mechanics." *International Journal of Engineering Science*, 3:533–537.

Grioli, G. (1960). "Elasticita asimmetrica." *Annali di Matematica Pura ed Applicata, Ser. IV.*, 50:389–417.

Günther, W. (1958). "Zur static und kinematik des kosseratishen kontinuums." *Abh. Braunschweig. Wiss. Ges.*, 10:195–213.

Guo, T. P., Nikolaev, P., Thess, A., Colbert, D. T., and Smalley, R. E. (1995). "Catalytic growth of single-walled nanotubes by laser vaporization." *Chemical Physics Letters*, 243:49–54.

Gurtin, M. E. (1965). "Thermodynamics and the possibility of spacial interactions in elastic materials." *Arch. Rat. Mech. Anal.*, 19:339–352.

Hampton, L. (1967). "Acoustic properties of sediments." *J. Acoust. Soc. Am.*, 42:882–890.

Haskell, N. A. (1953). "The dispersion of surface waves in multilayered media." *Bulletin Seismol. Soc. Amer.*, 43:17–34.

Haug, A. (1972). *Theoretical Solid State Physics*, Pergamon Press, New York. Two volumes.

Hiura, H., Ebbesen, T. W., Fujita, J., Tanigaki, K., and Takada, T. (1994). "Role of sp3 defect structures in graphite and carbon nanotubes." *Nature*, 367:148–151.

Hsu, W. K., Hare, J. P., Terrones, M., Kroto, H. W., Walton, D. R. M., and Harris, P. J. F. (1995). "Condensed-phase nanotubes." *Nature*, 377(26):687.

Hudson, J. A. (1980). *The Excitation and Propagation of Elastic Waves*, Cambridge University Press, Cambridge, England.

Hudson, J. A. (1982). "Overall properties of a cracked solid." *Proc. Cambridge Philosoph. Soc.*, 88:371–384.

Iida, K. (1938). "The velocity of elastic waves in sand." *Bull. Earthquake Res. Inst., Japan*, 16:131–144.

Iijima, S. and Ichihashi, T. (1993). "Single-shell carbon nanotubes of 1-nm diameter." *Nature*, 363:603–605.

Iijima, S. (1991). "Helical microtubules of graphitic carbon." *Nature*, 354:56–58.

Jaunzemis, W. (1967). *Continuum Mechanics*, Macmillan, New York.

Kestin, J. (1990). "A note on the relation between the hypothesis of local equilibrium and the Clausius-Duhem inequality." *J. Non-Equilib. Thermodyn.*, 15:193–212.

Kikuchi, N. and Oden, J. T. (1988). *Contact Problems in Elasticity: A Study of Variational Inequalities and Finite Element Methods*, SIAM, Philadelphia.

Kirkaldy, J. S. and Young, D. J. (1987). *Diffusion in the Condensed State*, The Institute of Metals, London.

Kittel, C. (1986). *Introduction to Solid State Physics*, John Wiley and Sons, Inc., New York, sixth edition.

Koiter, W. T. (1964). "Couple-stresses in the theory of elasticity, I and II." *Proc. Roy. Netherlands Acad. Sci.*, B67:17–44.

Kolsky, H. (1963). *Stress Waves in Solids*, Dover, New York.

Krishna Reddy, G. V. and Venkatasubramanian, N. K. (1978). "On the flexural rigidity of a micropolar elastic circular cylinder." *Journal of Applied Mechanics*, 45:429–431.

Kröner, E. (1967). "Elasticity theory of materials with long range cohesive forces." *International Journal of Solids and Structures*, 3:731–742.

Krumhansl, J. A. (1968). "Some considerations of the relations between solid state physics and generalized continuum mechanics." In Kröner, E., editor, *Proc. IUTAM Symposium Mechanics of Generalized Continua*, pages 298–311, Berlin, Springer-Verlag.

Kueny, A. and Grimsditch, M. (1982). "Surface waves in a layered media." *Phys. Review*, 26:4699–4702.

Kunin, I. A. (1982). *Elastic Media with Microstructure I. One-Dimensional Models*, volume 26 of *Springer Series in Solid-State Sciences*, Springer-Verlag, Berlin.

Kunin, I. A. (1983). *Elastic Media with Microstructure II. Three-Dimensional Models*, volume 44 of *Springer Series in Solid-State Sciences*, Springer-Verlag, Berlin.

Lakes, R. (1995). "Experimental methods for study of cosserat elastic solids and other generalized elastic continua." In Mühlhaus, H.-B., editor, *Continuum Models for Materials with Microstructure*, pages 1–25, New York, John Wiley.

Landau, L. D. and Lifshitz, E. M. (1959). *Theory of Elasticity*, Pergamon Press, London.

Landsberg, P. T. (1969). *Solid State Theory. Methods and Applications*, Wiley-Interscience, New York.

Leet, L. D. (1938). *Practical Seismology and Seismic Prospecting*, D.Appleton-Century Company, New York.

Li, E. and Bagster, D. F. (1990). "A new block model of heaps." *Powder Technology*, 63:277–283.

Li, E. and Bagster, D. F. (1993). "An idealized three-dimensional model of heaped granular materials." *Powder Technology*, 74:271–278.

Liffman, K., Chan, D. Y. C., and Hughes, B. D. (1992). "Force distribution in a two dimensional sand pile." *Powder Technology*, 72:255–267.

Loiseau, A., Willaime, F., Demoncy, N., Hug, G., and Pascard, H. (1996). "Boron nitride nanotubes with reduced number of layers synthesized by arc discharge." *Physical Review Letters*, 76(25):4737–4740.

Lord Rayleigh (1885). "On waves propagated along the plane surface of an elastic solid." *Proc., London Math. Soc.*, 17:4–11.

Love, A. E. H. (1911). *Some Problems of Geodynamics*, Cambridge University Press, London.

Love, A. E. H. (1944). *A Treatise on the Mathematical Theory of Elasticity*, Dover, New York.

Macelwane, J. B. (1949). *Introduction to Theoretical Seismology: Geodynamics*, St. Louis University, Missouri.

Maddalena, F. and Ferrari, M. (1995). "Viscoelasticity of granular materials." *Mechanics of Materials*, 20(3):241–250.

Madelung, O. (1978). *Introduction to Solid State Theory*, Springer-Verlag, Berlin.

Mal, A. K. and Singh, S. J. (1991). *Deformation of Elastic Solids*, Prentice Hall, Englewood Cliffs, New Jersey.

Malischewsky, P. (1987). *Surface Waves and Discontinuities*, Elsevier, Amsterdam.

Maradudin, A. A., Montroll, E. W., Weiss, G. H., and Ipatova, I. P. (1971). *Theory of Lattice Dynamics in the Harmonic Approximation*, Solid State Physics, Advances in Research and Application, Suppl. 3, Academic Press, New York, second edition.

Matsukawa, E. and Hunter, A. N. (1956). "The variation of sound velocity with stress in sand." *Proc. Phys. Soc., London, B*, 69:847–848.

Meek, J. W. and Wolf, J. P. (1993). "Why cone models can represent the elastic half-space." *Earthquake Engng. Struct. Dynamics*, 22:759–771.

Miklowitz, J. and Achenbach, J. D. (1978). *Modern Problems in Elastic Wave Propagation*, Wiley, New York.

Miklowitz, J. (1980). *The Theory of Waves and Wavegides*, North-Holland, Amsterdam.

Mindlin, R. D. and Tiersten, H. F. (1962). "Effects of couple-stresses in linear elasticity." *Arc. Rat. Mech. Anal.*, 11:415–448.

Mindlin, R. D. (1964). "Microstructure in linear elasticity." *Arch. Ratl. Mech. Analysis*, 16:51–78.

Mindlin, R. D. (1965). "Second gradient of strain and surface-tension in linear elasticity." *International Journal of Solids and Structures*, 1:417–438.

Mindlin, R. D. (1968). "Theories of elastic continua and crystal lattice theories." In Kröner, E., editor, *Proc., Symposium on Mechanics of Generalized Continua*, pages 312–320, Berlin, Springer-Verlag.

Mirsa, B. (1979). "Stress transmission in granular media." *J. Geotech. Engng. Div. Proc. ASCE*, 105:1101–1107.

Miyabe, N. (1935). "Notes on the block structure of the earth's crust." *Bulletin Earthquake Res. Inst., Tokyo Imperial Univ.*, 13:280–285.

Miyamoto, Y., Rubio, A., Cohen, M. L., and Louie, S. G. (1994). "Chiral tubules of hexagonal BC_2N." *Physical Review B*, 50:4976.

Naghdi, P. M. (1980). "On the role of the second law of thermodynamics in the mechanics of materials." *Energy*, 5:771–781.

Nikolaevskii, V. N. and Afanasiev, E. F. (1969). "On some examples of media with microstructure of continuous particles." *International Journal of Solids and Structures*, 5:671–678.

Noll, W. (1958). "A mathematical theory of the mechanical behavior of continuous media." *Arc. Rat. Mech. Anal.*, 2:197–226.

Nunziato, J. W. and Walsh, E. K. (1980). "On ideal multiphase mixtures with chemical reactions and diffusion." *Archive for Rational Mechanics and Analysis*, 73:285.

Oliner, A. A. (1978). *Acoustic Surface Waves*, volume 24 of *Topics in Applied Physics*, Springer-Verlag, Berlin.

Oliver, J. (1955). "Rayleigh waves on a cylindrical curved surface." *Earthquake Notes*, 26:24–25.

Ostoja-Starzewski, M. (1992). "Random fields and processes in mechanics of granular materials." In Shen, H. H., editor, *Advances in Micromechanics of Granular Materials: Proc., 2nd US/Japan Seminar on Micromechanics of Granular Materials*, pages 71–80, Amsterdam, Elsevier.

Ouverney, G., Zhong, W., and Tomanek, D. (1992). "Structural rigidity and low frequency vibrational modes of long carbon tubes." *Zeitschrift für Physik*, 27:93–96.

Passman, S. L. (1977). "Mixtures of granular materials." *International Journal of Engineering Science*, 15:117.

Pilant, W. L. (1979). *Elastic Waves in the Earth*, Elsevier, Amsterdam.

Pollitz, F. (1994). "Surface wave scattering from sharp lateral discontinuities." *J. Geophys. Res.*, 99:21891–21909.

Prakash, S. (1981). *Soil Dynamics*, McGraw-Hill, New York.

Rivlin, R. S. (1968). "Generalized mechanics of continuous media." In Kröner, E., editor, *Proc. IUTAM Symposium Mechanics of Generalized Continua*, pages 1–17, Berlin, Springer-Verlag.

Robinson, R. and Benites, R. (1995). "Synthetic seismicity models of multiple interacting faults." *J. Geophys. Res.*, 100:18229–18238.

Rollins, F. R., Lim, T. C., and Farnell, G. W. (1968). "Ultrasonic reflectivity and surface wave phenomena on surfaces of copper crystals." *Appl. Phys. Letters*, 12:236–238.

Rosenberg, H. M. (1975). *The Solid State: An Introduction to the Physics of Crystals for Students of Physics, Material Sciences and Engineering*, Oxford University Press, Oxford.

Royer, D. and Dieulesaint, E. (1984). "Rayleigh wave velocity and displacement in orthotropic, tetragonal, hexagonal and cubic crystals." *J. Acoust. Soc. Am.*, 76:1438–1444.

Ruoff, R. S., Tersoff, J., Lorents, D. C., Subramoney, S., and Chan, B. (1993). "Radial deformations of carbon nanotubes by van der Waals forces." *Nature*, 364:514–516.

Saito, R., Fujita, M., Dresselhaus, G., and Dresselhaus, M. S. (1992). "Electronic structure of chiral graphene tubules." *Applied Physics Letters*, 60(18):2204–2206.

Saito, M. (1967). "Excitation of free oscillations and surface waves by a point source in a vertically heterogeneous earth." *J. Geophys. Res.*, 72:3689–3699.

Sezawa, K. (1927). "Dispersion of elastic wave propagated on the surface of stratified bodies and on curved surfaces." *Bulletin Earthquake Res. Inst., Tokyo Imperial Univ.*, 3:1–18.

Shames, I. and Cozzarelli, F. (1991). *Elastic and Inelastic Stress Analysis*, Prentice-Hall, Englewood Cliffs, New Jersey.

Slade, R. E. and Walton, K. (1993). "Inter-granular friction and the mechanical and acoustic properties of granular media." In Thornton, C., editor, *Powders & Grains 93:Proc., 2nd Int. Conf. on Micromechanics of Granular Media*, pages 93–98, Rotterdam, ASMGM, A. A. Balkema.

Sokolnikoff, I. S. and Sokolnikoff, E. S. (1941). *Higher Mathematics for Engineers and Physicists*, McGraw-Hill, New York.

Stephan, O., Ajayan, P. M., Colliex, C., Redlich, P., Lambert, J. M., Bernier, P., and Lefin, P. (1994). "Doping graphitic and carbon nanotube structures with boron and nitrogen." *Science*, 266:1683–1685.

Stojanović, R. (1972). "On the mechanics of materials with microstructure." *Acta Mechanica*, 15:261–273.

Stoneley, R. (1955). "The propagation of the surface elastic waves in a cubic crystal." *Proc. Royal Soc. Acoust.*, 232:447–448.

Stout, R. B. (1989). "Statistical model for particle-void deformation kinetics in granular materials during shock wave propagation." In *Proc., Annual Meeting ASME Wave Propagation in Granular Media*, pages 51–74, New York, ASME.

Takahashi, T. and Sato, Y. (1949). "On the theory of elastic waves in granular substance." *Bull. Earthquake Res. Inst., Tokyo Univ.*, 27:11–16.

Takahashi, T. and Sato, Y. (1950). "On the theory of elastic waves in granular substance." *Bull. Earthquake Res. Inst., Tokyo Univ.*, 28:37–43.

Teodorescu, P. P. and Soós, E. (1973). "Discrete, quasi-continuous and continuous models of elastic solids." *Zeitschrift Angew. Math. Mech.*, 53:T33–T43.

Toda, M. (1988). *Theory of Nonlinear Lattices*, volume 20 of *Springer Series in Solid State Sciences*, Springer-Verlag, Berlin, second edition.

Todhunter, I. (1886). *A History of the Theory of Elasticity*, volume 1, Cambridge University Press.

Toupin, R. A. (1962). "Elastic materials with couple-stresses." *Arc. Rat. Mech. Anal.*, 11:385–414.

Toupin, R. A. (1964). "Theories of elasticity with couple-stress." *Arc. Rat. Mech. Anal.*, 17:85–112.

Toupin, R. A. (1968). "Dislocated and oriented media." In Kröner, E., editor, *Proc. IUTAM Symposium Mechanics of Generalized Continua*, pages 126–140, Berlin.

Trent, B. C. (1989). "Numerical simulation of wave propagation through cemented granular material." In Karamanlidis, D. and Stout, R. B., editors, *Proc., Annual Meeting ASME Wave Propagation in Granular Media*, pages 9–15, New York, ASME.

Truesdell, C. and Noll, W. (1965). *The Non-Linear Field Theories of Mechanics*, Handbuch der Physik, Springer-Verlag.

Truesdell, C. and Toupin, R. A. (1960). *Handbuch der Physik*, chapter The Classical Field Theories III/1. Springer-Verlag, Berlin.

Truesdell, C. (1957). *Rend. Lincei*, volume 22, .

Truesdell, C. (1969). *Rational Thermodynamics*, McGraw-Hill.

Urick, R. J. (1948). "The absorption of sound in suspensions of irregular particles." *J. Acoust. Soc. Am.*, 20:283–289.

Venkatamaran, G., Feldkamp, L. A., and Sahni, V. C. (1975). *Dynamics of Perfect Crystals*, MIT Press, Cambridge, MA.

Viktorov, I. (1967). *Rayleigh and Lamb Waves*, Plenum, New York.

Weng-Sieh, Z., Cherrey, K., Chopra, N. G., Blase, X., Miyamoto, Y., Rubio, A., Cohen, M. L., Gronsky, R., Louie, S. G., and Zettl, A. (1995). "Synthesis of $B_xC_yN_z$ nanotubes." *Physical Review B*, 51(16):11229–11232.

Wijesinghe, A. M. (1989). "A constitutive theory for stress wave propagation across closed fluid-filled rough fractures." In Karamanlidis, D. and Stout, R. B., editors, *Proc., Annual Meeting ASME Wave Propagation in Granular Media*, pages 75–90, New York, ASME.

Ziman, J. H. (1960). *Electrons and Phonons*, Oxford University Press, Oxford.

A. Micro-macromoduli Relations

K. Mon

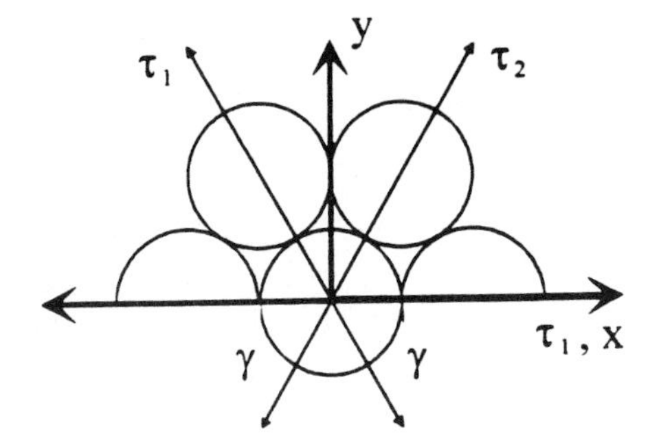

Fig. A.1. Planar $n = 3$ DM lattice.

The purpose of this appendix is to provide numeric values for the doublet mechanical micromoduli used in microconstitutive relations of the form

$$p_\alpha = \sum_{\beta=1}^{n} A_{\alpha\beta}\epsilon_\beta, \tag{A.1}$$

where $A_{\alpha\beta}$ is a matrix of micromoduli. Use is made of the isotropic $n = 3$ doublet planar hexagonal ($\gamma = 60°$) lattice (Fig. A.1), and the micro to macro moduli relation:

$$\hat{\mathbf{C}} = \mathbf{M}\mathbf{A}\mathbf{M}^{\mathrm{T}} \tag{A.2}$$

defined in eqn. (8.22). In terms of the Lamé constants, λ and μ, the form of $\hat{\mathbf{C}}$ appropriate for a macroscopically isotropic material undergoing plane strain is

$$\hat{\mathbf{C}} = \begin{bmatrix} \lambda + 2\mu & \lambda & 0 \\ \lambda & \lambda + 2\mu & 0 \\ 0 & 0 & \mu \end{bmatrix}, \tag{A.3}$$

such that macroscopically the in-plane stress-strain relation is

$$\hat{\sigma} = \hat{\mathbf{C}}\hat{\epsilon}, \tag{A.4}$$

where $\hat{\sigma} := \{\sigma_{11}, \sigma_{22}, \sigma_{12}\}^T$ and $\hat{\epsilon} := \{\epsilon_{11}, \epsilon_{22}, \epsilon_{12} + \epsilon_{21}\}^T$. The form of $\hat{\mathbf{C}}$ appropriate for a macroscopically isotropic material undergoing plane stress, which we elect to call $\hat{\mathbf{Q}}$, is

$$\hat{\mathbf{Q}} = \begin{bmatrix} 4\mu\left(\frac{\lambda+\mu}{\lambda+2\mu}\right) & \left(\frac{2\lambda\mu}{\lambda+2\mu}\right) & 0 \\ \left(\frac{2\lambda\mu}{\lambda+2\mu}\right) & 4\mu\left(\frac{\lambda+\mu}{\lambda+2\mu}\right) & 0 \\ 0 & 0 & \mu \end{bmatrix}, \tag{A.5}$$

such that macroscopically the in-plane stress-strain relation is

$$\hat{\sigma} = \hat{\mathbf{Q}}\hat{\epsilon}, \tag{A.6}$$

and we define a micromoduli matrix $B_{\alpha\beta}$, analogous to $A_{\alpha\beta}$ in eqn. (A.1), by the relations

$$\hat{\mathbf{Q}} = \mathbf{MBM^T}, \tag{A.7}$$

$$p_\alpha = \sum_{\beta=1}^{n} B_{\alpha\beta}\,\epsilon_\beta. \tag{A.8}$$

The matrices $A_{\alpha\beta}$ and $B_{\alpha\beta}$ obtained from eqns. (A.2) and (A.7) using macromoduli matrices (A.3) and (A.5) are both of the form

$$\begin{bmatrix} a & b & b \\ b & a & b \\ b & b & a \end{bmatrix}. \tag{A.9}$$

and are independent of in plane rotations of the DM lattice vectors ($\bar{\tau}_\alpha$ in Fig. A.1) relative to the macroscopic coordinate system (x and y in Fig. A.1). Relation (A.9) is also in complete agreement with the constitutive restrictions obtained on DM linear elasticity in Chaps. 2 and 3. Thus, for each type of planar problem, only two micromoduli (a and b) are needed. For the case of plane strain, the components of $A_{\alpha\beta}$ are such that

$$a = \frac{4}{9}(\lambda + 5\mu), \tag{A.10}$$

$$b = \frac{4}{9}(\lambda - \mu), \tag{A.11}$$

while for plane stress, the components of $B_{\alpha\beta}$ are such that

$$a = \frac{4}{9}\mu\left(\frac{7\lambda + 10\mu}{\lambda + 2\mu}\right), \tag{A.12}$$

$$b = \frac{4}{9}\mu\left(\frac{\lambda - 2\mu}{\lambda + 2\mu}\right). \tag{A.13}$$

Table A.1 provides a listing of values for a and b for a variety of materials.

One could use the ratio a/b as a quantitative guide to the applicability of the more simple constitutive relations

$$p_\alpha = A_o \epsilon_\alpha, \tag{A.14}$$

$$p_\alpha = B_o \epsilon_\alpha, \tag{A.15}$$

using eqn. (A.14) or eqn. (A.15) instead of eqn. (A.1) or (A.8) only when $a/b \gg 1$ (usually when $\nu \cong 1/4$ or $\lambda \cong \mu$ for plane strain, $\nu \cong 1/3$ or $\lambda \cong 2\mu$ for plane stress).

Table A.1. Isotropic micromoduli for plane strain ($A_{\alpha\beta}$) and plane stress ($B_{\alpha\beta}$) problems with macroscopic isotropy and the planar hexagonal DM lattice.

Material	E (GPa)	ν	$A_{\alpha\beta}$ (GPa)		$B_{\alpha\beta}$ (GPa)	
			a	b	a	b
Ag	82.7	0.367	104.3	23.7	82.8	2.1
Al	70.3	0.345	83.9	14.2	70.3	0.62
Au	78.0	0.44	148.5	76.2	79.1	6.9
Bi	31.9	0.33	37.0	5.0	31.9	– 0.08
Cd	49.9	0.30	55.4	4.3	50.0	– 1.2
Cr	279.1	0.21	293.4	– 14.1	283.5	– 24.1
Cu	129.8	0.343	154.3	25.4	129.8	0.95
Fe (cast)	152.3	0.27	164.5	4.6	153.0	– 6.9
Mg	44.7	0.29	149.2	3.0	44.8	– 1.4
Mo	275.8	0.32	314.7	36.1	275.9	– 2.7
Ni (unmag.)	219.2	0.306	245.3	21.5	219.4	– 4.4
Pb	16.1	0.44	30.6	15.7	16.3	1.4
Pt	168	0.377	218.7	56.0	168.4	5.7
Pu	96.5	0.18	101.1	– 8.0	98.9	– 10.2
Sn	49.9	0.357	61.3	12.2	49.9	0.90
Ta	185.7	0.342	220.3	35.8	185.7	1.2
Th	58.6	0.27	63.3	1.8	58.9	– 2.7
Ti	115.7	0.321	132.2	15.4	115.7	– 1.1
U	165.5	0.21	174.0	– 8.4	168.1	– 14.2
V	127.6	0.365	160.0	35.4	127.7	3.1
W	411	0.28	447.6	19.5	412.3	– 15.9
Zn	108.4	0.249	115.6	– 0.15	109.2	– 6.5
Brass (Zn – 30 Cu)	100.6	0.35	121.4	22.1	100.6	1.3
Constantan	162.4	0.327	187.4	24.2	162.4	– 0.77
Invar	144	0.259	154.4	1.9	144.8	–7.6
Mild Steel	211.9	0.291	233.2	14.3	212.3	–6.5
Stainless Steel	215.3	0.283	235.1	11.3	215.9	–7.9
WC	534.4	0.22	563.2	– 20.9	541.6	– 42.4
Crown Glass	71.3	0.22	75.1	– 2.8	72.3	–5.7
Fused Quartz	73.1	0.17	76.6	– 6.7	75.1	– 8.2
Heavy Flint Glass	80.1	0.27	86.5	2.4	80.4	– 3.6
Polyamide	2.07	0.4	3.0	1.0	2.1	0.1
Polyethylene	0.69	0.46	1.7	1.1	0.70	0.07

[1] James, A. and Lord, M. (1992), Index of Chemical and Physical Data, MacMillan Press, Ltd., New York, New York, p. 20-21.
[2] Handbook of Tables for Applied Engineering Science, 2nd Ed. (1973), CRC Press, Inc., Boca Raton, Florida, pp. 68, 117.

Lecture Notes in Physics

For information about Vols. 1–444
please contact your bookseller or Springer-Verlag

Vol. 445: J. Brey, J. Marro, J. M. Rubí, M. San Miguel (Eds.), 25 Years of Non-Equilibrium Statistical Mechanics. Proceedings, 1994. XVII, 387 pages. 1995.

Vol. 446: V. Rivasseau (Ed.), Constructive Physics. Results in Field Theory, Statistical Mechanics and Condensed Matter Physics. Proceedings, 1994. X, 337 pages. 1995.

Vol. 447: G. Aktaş, C. Saçlıoğlu, M. Serdaroğlu (Eds.), Strings and Symmetries. Proceedings, 1994. XIV, 389 pages. 1995.

Vol. 448: P. L. Garrido, J. Marro (Eds.), Third Granada Lectures in Computational Physics. Proceedings, 1994. XIV, 346 pages. 1995.

Vol. 449: J. Buckmaster, T. Takeno (Eds.), Modeling in Combustion Science. Proceedings, 1994. X, 369 pages. 1995.

Vol. 450: M. F. Shlesinger, G. M. Zaslavsky, U. Frisch (Eds.), Lévy Flights and Related Topics in Physics. Proceedigs, 1994. XIV, 347 pages. 1995.

Vol. 451: P. Krée, W. Wedig (Eds.), Probabilistic Methods in Applied Physics. IX, 393 pages. 1995.

Vol. 452: A. M. Bernstein, B. R. Holstein (Eds.), Chiral Dynamics: Theory and Experiment. Proceedings, 1994. VIII, 351 pages. 1995.

Vol. 453: S. M. Deshpande, S. S. Desai, R. Narasimha (Eds.), Fourteenth International Conference on Numerical Methods in Fluid Dynamics. Proceedings, 1994. XIII, 589 pages. 1995.

Vol. 454: J. Greiner, H. W. Duerbeck, R. E. Gershberg (Eds.), Flares and Flashes, Germany 1994. XXII, 477 pages. 1995.

Vol. 455: F. Occhionero (Ed.), Birth of the Universe and Fundamental Physics. Proceedings, 1994. XV, 387 pages. 1995.

Vol. 456: H. B. Geyer (Ed.), Field Theory, Topology and Condensed Matter Physics. Proceedings, 1994. XII, 206 pages. 1995.

Vol. 457: P. Garbaczewski, M. Wolf, A. Weron (Eds.), Chaos – The Interplay Between Stochastic and Deterministic Behaviour. Proceedings, 1995. XII, 573 pages. 1995.

Vol. 458: I. W. Roxburgh, J.-L. Masnou (Eds.), Physical Processes in Astrophysics. Proceedings, 1993. XII, 249 pages. 1995.

Vol. 459: G. Winnewisser, G. C. Pelz (Eds.), The Physics and Chemistry of Interstellar Molecular Clouds. Proceedings, 1993. XV, 393 pages. 1995.

Vol. 460: S. Cotsakis, G. W. Gibbons (Eds.), Global Structure and Evolution in General Relativity. Proceedings, 1994. IX, 173 pages. 1996.

Vol. 461: R. López-Peña, R. Capovilla, R. García-Pelayo, H. Waelbroeck, F. Zertuche (Eds.), Complex Systems and Binary Networks. Lectures, México 1995. X, 223 pages. 1995.

Vol. 462: M. Meneguzzi, A. Pouquet, P.-L. Sulem (Eds.), Small-Scale Structures in Three-Dimensional Hydrodynamic and Magnetohydrodynamic Turbulence. Proceedings, 1995. IX, 421 pages. 1995.

Vol. 463: H. Hippelein, K. Meisenheimer, H.-J. Röser (Eds.), Galaxies in the Young Universe. Proceedings, 1994. XV, 314 pages. 1995.

Vol. 464: L. Ratke, H. U. Walter, B. Feuerbach (Eds.), Materials and Fluids Under Low Gravity. Proceedings, 1994. XVIII, 424 pages, 1996.

Vol. 465: S. Beckwith, J. Staude, A. Quetz, A. Natta (Eds.), Disks and Outflows Around Young Stars. Proceedings, 1994. XII, 361 pages, 1996.

Vol. 466: H. Ebert, G. Schütz (Eds.), Spin – Orbit-Influenced Spectroscopies of Magnetic Solids. Proceedings, 1995. VII, 287 pages, 1996.

Vol. 467: A. Steinchen (Ed.), Dynamics of Multiphase Flows Across Interfaces. 1994/1995. XII, 267 pages. 1996.

Vol. 468: C. Chiuderi, G. Einaudi (Eds.), Plasma Astrophysics. 1994. VII, 326 pages. 1996.

Vol. 469: H. Grosse, L. Pittner (Eds.), Low-Dimensional Models in Statistical Physics and Quantum Field Theory. Proceedings, 1995. XVII, 339 pages. 1996.

Vol. 470: E. Martínez-González, J. L. Sanz (Eds.), The Universe at High-z, Large-Scale Structure and the Cosmic Microwave Background. Proceedings, 1995. VIII, 254 pages. 1996.

Vol. 471: W. Kundt (Ed.), Jets from Stars and Galactic Nuclei. Proceedings, 1995. X, 290 pages. 1996.

Vol. 472: J. Greiner (Ed.), Supersoft X-Ray Sources. Proceedings, 1996. XIII, 350 pages. 1996.

Vol. 473: P. Weingartner, G. Schurz (Eds.), Law and Prediction in the Light of Chaos Research. X, 291 pages. 1996.

Vol. 474: Aa. Sandqvist, P. O. Lindblad (Eds.), Barred Galaxies and Circumnuclear Activity. Proceedings of the Nobel Symposium 98, 1995. XI, 306 pages. 1996.

Vol. 475: J. Klamut, B. W. Veal, B. M. Dabrowski, P. W. Klamut, M. Kazimierski (Eds.), Recent Developments in High Temperature Superconductivity. Proceedings, 1995. XIII, 362 pages. 1996.

Vol. 476: J. Parisi, S. C. Müller, W. Zimmermann (Eds.), Nonlinear Physics of Complex Systems. Current Status and Future Trends. XIII, 388 pages. 1996.

Vol. 477: Z. Petru, J. Przystawa, K. Rapcewicz (Eds.), From Quantum Mechanics to Technology. Proceedings, 1996. IX, 379 pages. 1996.

New Series m: Monographs

Vol. m 1: H. Hora, Plasmas at High Temperature and Density. VIII, 442 pages. 1991.

Vol. m 2: P. Busch, P. J. Lahti, P. Mittelstaedt, The Quantum Theory of Measurement. XIII, 165 pages. 1991. Second Revised Edition: XIII, 181 pages. 1996.

Vol. m 3: A. Heck, J. M. Perdang (Eds.), Applying Fractals in Astronomy. IX, 210 pages. 1991.

Vol. m 4: R. K. Zeytounian, Mécanique des fluides fondamentale. XV, 615 pages, 1991.

Vol. m 5: R. K. Zeytounian, Meteorological Fluid Dynamics. XI, 346 pages. 1991.

Vol. m 6: N. M. J. Woodhouse, Special Relativity. VIII, 86 pages. 1992.

Vol. m 7: G. Morandi, The Role of Topology in Classical and Quantum Physics. XIII, 239 pages. 1992.

Vol. m 8: D. Funaro, Polynomial Approximation of Differential Equations. X, 305 pages. 1992.

Vol. m 9: M. Namiki, Stochastic Quantization. X, 217 pages. 1992.

Vol. m 10: J. Hoppe, Lectures on Integrable Systems. VII, 111 pages. 1992.

Vol. m 11: A. D. Yaghjian, Relativistic Dynamics of a Charged Sphere. XII, 115 pages. 1992.

Vol. m 12: G. Esposito, Quantum Gravity, Quantum Cosmology and Lorentzian Geometries. Second Corrected and Enlarged Edition. XVIII, 349 pages. 1994.

Vol. m 13: M. Klein, A. Knauf, Classical Planar Scattering by Coulombic Potentials. V, 142 pages. 1992.

Vol. m 14: A. Lerda, Anyons. XI, 138 pages. 1992.

Vol. m 15: N. Peters, B. Rogg (Eds.), Reduced Kinetic Mechanisms for Applications in Combustion Systems. X, 360 pages. 1993.

Vol. m 16: P. Christe, M. Henkel, Introduction to Conformal Invariance and Its Applications to Critical Phenomena. XV, 260 pages. 1993.

Vol. m 17: M. Schoen, Computer Simulation of Condensed Phases in Complex Geometries. X, 136 pages. 1993.

Vol. m 18: H. Carmichael, An Open Systems Approach to Quantum Optics. X, 179 pages. 1993.

Vol. m 19: S. D. Bogan, M. K. Hinders, Interface Effects in Elastic Wave Scattering. XII, 182 pages. 1994.

Vol. m 20: E. Abdalla, M. C. B. Abdalla, D. Dalmazi, A. Zadra, 2D-Gravity in Non-Critical Strings. IX, 319 pages. 1994.

Vol. m 21: G. P. Berman, E. N. Bulgakov, D. D. Holm, Crossover-Time in Quantum Boson and Spin Systems. XI, 268 pages. 1994.

Vol. m 22: M.-O. Hongler, Chaotic and Stochastic Behaviour in Automatic Production Lines. V, 85 pages. 1994.

Vol. m 23: V. S. Viswanath, G. Müller, The Recursion Method. X, 259 pages. 1994.

Vol. m 24: A. Ern, V. Giovangigli, Multicomponent Transport Algorithms. XIV, 427 pages. 1994.

Vol. m 25: A. V. Bogdanov, G. V. Dubrovskiy, M. P. Krutikov, D. V. Kulginov, V. M. Strelchenya, Interaction of Gases with Surfaces. XIV, 132 pages. 1995.

Vol. m 26: M. Dineykhan, G. V. Efimov, G. Ganbold, S. N. Nedelko, Oscillator Representation in Quantum Physics. IX, 279 pages. 1995.

Vol. m 27: J. T. Ottesen, Infinite Dimensional Groups and Algebras in Quantum Physics. IX, 218 pages. 1995.

Vol. m 28: O. Piguet, S. P. Sorella, Algebraic Renormalization. IX, 134 pages. 1995.

Vol. m 29: C. Bendjaballah, Introduction to Photon Communication. VII, 193 pages. 1995.

Vol. m 30: A. J. Greer, W. J. Kossler, Low Magnetic Fields in Anisotropic Superconductors. VII, 161 pages. 1995.

Vol. m 31: P. Busch, M. Grabowski, P. J. Lahti, Operational Quantum Physics. XI, 230 pages. 1995.

Vol. m 32: L. de Broglie, Diverses questions de mécanique et de thermodynamique classiques et relativistes. XII, 198 pages. 1995.

Vol. m 33: R. Alkofer, H. Reinhardt, Chiral Quark Dynamics. VIII, 115 pages. 1995.

Vol. m 34: R. Jost, Das Märchen vom Elfenbeinernen Turm. VIII, 286 pages. 1995.

Vol. m 35: E. Elizalde, Ten Physical Applications of Spectral Zeta Functions. XIV, 228 pages. 1995.

Vol. m 36: G. Dunne, Self-Dual Chern-Simons Theories. X, 217 pages. 1995.

Vol. m 37: S. Childress, A.D. Gilbert, Stretch, Twist, Fold: The Fast Dynamo. XI, 410 pages. 1995.

Vol. m 38: J. González, M. A. Martín-Delgado, G. Sierra, A. H. Vozmediano, Quantum Electron Liquids and High-T_c Superconductivity. X, 299 pages. 1995.

Vol. m 39: L. Pittner, Algebraic Foundations of Non-Commutative Differential Geometry and Quantum Groups. XII, 469 pages. 1996.

Vol. m 40: H.-J. Borchers, Translation Group and Particle Representations in Quantum Field Theory. VII, 131 pages. 1996.

Vol. m 41: B. K. Chakrabarti, A. Dutta, P. Sen, Quantum Ising Phases and Transitions in Transverse Ising Models. X, 204 pages. 1996.

Vol. m 42: P. Bouwknegt, J. McCarthy, K. Pilch, The $\mathcal{W}_3$ Algebra. Modules, Semi-infinite Cohomology and BV Algebras. XI, 204 pages. 1996.

Vol. m 43:

Vol. m 44: A. Bach, Indistinguishable Classical Particles. VIII, 157 pages. 1997.

Vol. m 45: M. Ferrari, V. T. Granik, A. Imam, J. C. Nadeau (Eds.), Advances in Doublet Mechanics. XVI, 214 pages. 1997.